Reefs at Risk Revisited

LAURETTA BURKE | KATHLEEN REYTAR
MARK SPALDING | ALLISON PERRY

Contributing Authors

Emily Cooper, Benjamin Kushner,
Elizabeth Selig, Benjamin Starkhouse,
Kristian Teleki, Richard Waite,
Clive Wilkinson, Terri Young

WORLD
RESOURCES
INSTITUTE

WASHINGTON, DC

Hyacinth Billings
Publications Director

Cover Photo
Tornado of Fish by Michael Emerson

Inside Front Cover Photo
Suchana Chavanich/Marine Photobank

Layout of *Reefs at Risk Revisited*
Maggie Powell

Each World Resources Institute report represents a timely, scholarly treatment of a subject of public concern. WRI takes responsibility for choosing the study topics and guaranteeing its authors and researchers freedom of inquiry. It also solicits and responds to the guidance of advisory panels and expert reviewers. Unless otherwise stated, however, all the interpretation and findings set forth in WRI publications are those of the authors.

ISBN 978-1-56973-762-0
Library of Congress Control Number: 2011922172

Mixed Sources
Product group from well-managed forests, controlled sources and recycled wood or fiber
FSC
www.fsc.org Cert. no. SGS-COC-003732
© 1996 Forest Stewardship Council

Printed on FSC certified paper, produced using 100% Certified Renewable Energy.

Contents

MAPS

TABLES

FIGURES

Foreword

As anyone who has spent time around the ocean knows—whether diving, conducting research, or fishing—coral reefs are among the world's greatest sources of beauty and wonder. Home to over 4,000 species of fish and 800 types of coral, reefs offer an amazing panorama of underwater life.

Coral reefs supply a wide range of important benefits to communities around the world. From the fisherman in Indonesia or Tanzania who relies on local fish to feed his family, to the scientist in Panama who investigates the medicinal potential of reef-related compounds, reefs provide jobs, livelihoods, food, shelter, and protection for coastal communities and the shorelines along which they live.

Unfortunately, reefs today are facing multiple threats from many directions. 2010 was one of the warmest years on record, causing widespread damage to coral reefs. Warmer oceans lead to coral bleaching, which is becoming increasingly frequent around the globe—leaving reefs, fish, and the communities who depend on these resources at great risk. No one yet knows what the long-term impacts of this bleaching will be. But, if the ocean's waters keep warming, the outlook is grim.

Against this backdrop, the World Resources Institute has produced *Reefs at Risk Revisited,* a groundbreaking new analysis of threats to the world's coral reefs. This report builds on WRI's seminal 1998 report, *Reefs at Risk,* which served as a call to action for policymakers, scientists, nongovernmental organizations, and industry to confront one of the most pressing, though poorly understood, environmental issues. That report played a critical role in raising awareness and driving action, inspiring countless regional projects, stimulating greater funding, and providing motivation for new policies to protect marine areas and mitigate risks.

However, much has changed since 1998—including an increase in the world's population, and with it greater consumption, trade, and tourism. Rising economies in the developing world have led to more industrialization, more agricultural development, more commerce, and more and more greenhouse gas emissions. All of these factors have contributed to the need to update and refine the earlier report.

The latest report builds on the original *Reefs at Risk* in two important ways. First, the map-based assessment uses the latest global data and satellite imagery, drawing on a reef map that is 64 times more detailed than in the 1998 report. The second major new component is our greater understanding of the effects of climate change on coral reefs. As harmful as overfishing, coastal development, and other local threats are to reefs, the warming planet is quickly becoming the chief threat to the health of coral reefs around the world. Every day, we dump 90 million tons of carbon pollution into the thin shell of atmosphere surrounding our planet—roughly one-third of it goes into the ocean, increasing ocean acidification.

Coral reefs are harbingers of change. Like the proverbial "canary in the coal mine," the degradation of coral reefs is a clear sign that our dangerous overreliance on fossil fuels is already changing Earth's climate. Coral reefs are currently experiencing higher ocean temperatures and acidity than at any other time in at least the last 400,000 years. If we continue down this path, all corals will likely be threatened by mid-century, with 75 percent facing high to critical threat levels.

Reefs at Risk Revisited reveals a new reality about coral reefs and the increasing stresses they are under. It should serve as a wake-up call for policymakers and citizens around the world. By nature, coral reefs have proven to be resilient and can bounce back from the effects of a particular threat. But, if we fail to address the multiple threats they face, we will likely see these precious ecosystems unravel, and with them the numerous benefits that people around the globe derive from these ecological wonders. We simply cannot afford to let that happen.

HON. AL GORE

Former Vice President of the United States

Abbreviations and Acronyms

AGRRA	The Atlantic and Gulf Rapid Reef Assessment Program
AIMS	The Australian Institute of Marine Science
CITES	Convention on International Trade in Endangered Species
CO_2	Carbon dioxide
COTS	Crown-of-thorns starfish
CORDIO	Coastal Oceans Research and Development in the Indian Ocean
DHW	Degree heating week
FAO	Food and Agriculture Organization of the United Nations
CRIOBE	Le Centre de Recherches Insulaires et Observatoire de l'Environnement
GCRMN	The Global Coral Reef Monitoring Network
GDP	Gross domestic product
GIS	Geographic Information System
ICRAN	International Coral Reef Action Network
ICRI	International Coral Reef Initiative
IMaRS/USF	Institute for Marine Remote Sensing, University of South Florida
IMO	International Maritime Organization
IPCC	Intergovernmental Panel on Climate Change
IRD	Institut de Recherche pour le Développement
IUCN	International Union for Conservation of Nature
LDC	Least developed country
LMMAs	Locally managed marine areas
MAC	Marine Aquarium Council
MPAs	Marine protected areas
MARPOL	International Convention for the Prevention of Pollution from Ships
NASA	U.S. National Aeronautics and Space Administration
NGOs	Nongovernmental organizations
NOAA	U.S. National Oceanic and Atmospheric Administration
OPRC	International Convention on Oil Pollution Preparedness, Response, and Cooperation
PICRC	Palau International Coral Reef Center
ppm	Parts Per Million
REEF	The Reef Environmental Education Foundation
sq km	Square kilometers
SST	Sea surface temperature
TNC	The Nature Conservancy
UNEP-WCMC	United Nations Environment Programme-World Conservation Monitoring Centre
UNFCCC	United Nations Framework Convention on Climate Change
WCS	Wildlife Conservation Society
WDPA	World Database of Protected Areas
WRI	World Resources Institute
WWF	World Wildlife Fund

PHOTO: UWE DEICHMANN

Acknowledgments

The *Reefs at Risk Revisited* project would not have been possible without the encouragement and financial support provided by The Roy Disney Family Foundation, The David and Lucile Packard Foundation, The Marisla Foundation, Netherlands Ministry of Foreign Affairs, The Chino Cienega Foundation, U.S. Department of the Interior, U.S. Department of State through the International Coral Reef Initiative, U.S. National Oceanic and Atmospheric Administration through the National Fish and Wildlife Foundation, The Tiffany & Co. Foundation, The Henry Foundation, The Ocean Foundation, Project AWARE Foundation, and The Nature Conservancy.

Reefs at Risk Revisited is the result of a more than two-year effort, involving a broad network of partners. The World Resources Institute gratefully acknowledges the many partners and colleagues who contributed to this project. (See inside front cover for full institutional names.) Clive Wilkinson (GCRMN) synthesized information from the GCRMN network on coral reef status and trends; Kristian Teleki (SeaWeb) served as catalyst and sounding board throughout the project; Terri Young (ICRAN) led the effort to collect and edit reef stories for the report; Richard Waite and Benjamin Kushner (WRI) provided broad support on research, editing, fundraising, communication, and reef stories; Elizabeth Selig (CI) provided valuable guidance on how to incorporate climate change and ocean acidification into the threat analysis; Emily Cooper (ERM) provided guidance on the social and economic contribution of coral reefs; Benjamin Starkhouse (WorldFish Center) provided essential research support for the social vulnerability analysis; and Moi Khim Tan (WorldFish Center) was a tireless provider of high quality data on coral reef locations, MPAs, coral bleaching, and disease through the ReefBase network.

Reefs at Risk Revisited relies on spatial and statistical data from a wide range of sources, coupled with expert guidance on data integration, modeling methods, and an extensive review of model results.

Coral Reef Map. Many partners contributed to the global coral reef map used in this analysis. The map is based on data from the Millennium Coral Reef Mapping Project of the Institute for Marine Remote Sensing, University of South Florida (IMaRS/USF), and the Institute de Recherche Pour le Developpement (IRD), with funding from NASA; thanks to Serge Andréfouët (IRD), Julie Robinson (NASA), and Frank Muller-Karger (USF). These data were augmented with data collected over many years at UNEP-WCMC, as well as data provided by Stacy Jupiter and Ingrid Pifeleti Qauqau (WCS-Fiji), Jamie Monty (Florida Dept. of Environmental Protection), Ashraf Saad Al-Cibahy (Environment Agency, Abu Dhabi), Helena Pavese (UNEP-WCMC) , David Holst (Great Barrier Reef Marine Park Authority), Eduardo Klein (Institute of Technology and Marine Sciences, Venezuela), and NOAA. Data were meticulously edited and integrated by Corinna Ravilious (UNEP-WCMC) and Moi Khim Tan (WorldFish Center).

Coral Condition, Bleaching, and Disease. Incorporating data on coral condition from surveys and observations is an important component of the analysis for both assessing trends over time and calibrating model results. Bleaching and disease data were provided by ReefBase, WorldFish Center, and UNEP-WCMC. Providers of monitoring and assessment data used in the analysis included Reef Check, AGRRA, and GCRMN (see Appendix 2 for additional details). Partners who contributed to this component include Gregor Hodgson and Jenny Mihaly (Reef Check); Laurie Raymundo (University of Guam); Caroline Rogers (U.S. Geological Survey); Melanie McField (Smithsonian Institution); Judy Lang (Independent); Robert Ginsburg (University of Miami); Jos Hill (Reef Check Australia), Enric Sala (National Geographic), and Jeffrey Wielgus (WRI).

Changing Climate. Ocean warming and ocean acidification were not included in previous *Reefs at Risk* analyses. Many partners contributed to this enhancement. Tyler Christensen and Mark Eakin (NOAA Coral Reef Watch) and Ken Casey and Tess Brandon (NOAA Oceanographic Data Center) provided data on past thermal stress. Simon Donner (University of British Columbia) provided projections of future thermal stress. Long Cao and Ken Caldeira (Stanford University) provided data-modeled estimates of present and future aragonite saturation state (ocean acidification). Many people provided information on these threats, advice on modeling, or reviewed results, including Bob Buddemeier (Kansas Geological Survey); Gabriel Grimsditch and David Obura (IUCN); Ellycia Harrould-Kolieb (Oceana); Joan Kleypas (National Center for Atmospheric Research); Nancy Knowlton (Smithsonian Institution); Jonathan Kool (James Cook University); Joseph Maina and Tim McClanahan (WCS); Rod Salm and Elizabeth McLeod (TNC); Peter Mumby (University of Queensland); Paul Marshall (Great Barrier Reef Marine Park Authority); and Terry Done (Australian Institute of Marine Science).

Watershed-based Pollution. Ben Halpern (NCEAS), Shaun Walbridge (UCSB), Michelle Devlin (James Cook University), Carmen Revenga (TNC), and Bart Wickel (WWF) provided assistance with data and modeling on erosion, sediment transport, and plume modeling for watershed-based pollution.

Overfishing and Destructive Fishing. Many people provided input on the analysis of overfishing and destructive fishing, including Elodie Lagouy (Reef Check Polynesia); Christy Semmens (REEF); Hugh Govan (LMMA Network); Annick Cros, Alan White, Arief Darmawan, Eleanor Carter, and Andreas Muljadi (TNC); Ken Kassem, Sikula Magupin, Cathy Plume, Helen Fox, Chrisma Salao, and Lida Pet-Soede (WWF); Melita Samoilys (CORDIO); Rick MacPherson (CORAL); Ficar Mochtar (Destructive Fishing Watch Indonesia); Daniel Ponce-Taylor and Monique Mancilla (Global Vision International); Patrick Mesia (Solomon Islands Dept. of Fisheries); and the Tanzania Dynamite Fishing Monitoring Network, especially Sibylle Riedmiller (coordinator), Jason Rubens, Lindsey West, Matt Richmond, Farhat Jah, Charles Dobie, Brian Stanley-Jackson, Isobel Pring, and John Van der Loon.

Social and Economic Vulnerability. This is the first global-scale assessment of vulnerability to reef loss ever completed. We are grateful to Christy Loper (NOAA); Nick Dulvy (SFU); Norbert Henninger (WRI); and David Mills, Edward Allison, Marie-Caroline Badjeck, Neil Andrew, and Diemuth Pemsl (WorldFish Center) for advising on methods, assisting with indicator development, and general support. Many people helpfully provided information or clarification on aspects of reef dependence and adaptive capacity, including Dedi Adhuri, Md. Giasuddin Khan, and William Collis (WorldFish Center); Andrea Coloma (Machalilla National Park, Ecuador); Roeland Dreischor (Central Bureau of Statistics, Netherlands Antilles); Andrew Gibbs (Australian Bureau of Statistics); James Gumbs and Stuart Wynne (Dept. of Fisheries and Marine Resources, Anguilla); Ian Horsford (Fisheries Division, Antigua and Barbuda); Ayana Johnson (Scripps Institution of Oceanography); Scotty Kyle (Ezemvelo KwaZulu-Natal Wildlife); Albert Leung (HK Agriculture, Fisheries and Conservation Department); Upul Liyanage (NARA, Sri Lanka); Kathy Lockhart (Dept. of Environment and Coastal Resources, Turks and Caicos); Hongguang Ma and Stewart Allen (PIFSC-NOAA); Weimin Miao (FAO-RAP); Nicholas Paul (James Cook University); Hideo Sekiguchi (Mie University); Elizabeth Taylor Jay (CORALINA); Qui Yongsong (South China Sea Fisheries Research Institute); and Jeffery Wielgus (WRI). Several people also provided data and other helpful support in developing economic and tourism indicators, including Giulia Carbone (IUCN); Azucena Pernia and Luigi Cabrini (World Tourism Organization); Jenny Miller-Garmendia (Project AWARE); and Mary Simon and Nathan Vizcarra (PADI).

Marine Protected Areas. Over many years, many people have provided valuable input regarding protected areas data from the World Database on Protected Areas, enhanced with inputs from ReefBase, the WorldFish Center, and TNC Indonesia. The development and application of the effectiveness scoring was begun with the Regional Reefs at Risk assessments, with considerable expert advice and data input. For the present study we are very grateful to the following for their assistance in reviewing MPA information, developing or correcting the methods, scoring MPA effectiveness, or commenting on the whole process: Venetia Hargreaves-Allen (University of London); Hugh Govan (LMMA Network); Abdul Halim, Alan White, Sangeeta Mangubhai, Steve Schill, Stuart Sheppard, John Knowles, and Juan Bezaury (TNC); Arjan Rajasuriya (GCRMN National Coordinator for Sri Lanka and National Aquatic Resources Research and Development Agency); Bruce Cauvin (GCRMN Regional Coordinator for Southwest Indian Ocean Islands); Camilo Mora (Dalhousie

University); John Day (GBRMPA); Heidi Schuttenberg (CSIRO, Australia); Jenny Waddell (NOAA); Dan Laffoley and Caitlin Toropova (IUCN/World Commission on Protected Areas); Peyman Eghtesadi (GCRMN Regional Coordinator, ROMPE and Iranian National Center for Oceanography); Mohammad Reza Fatemi (Islamic Azad University, Iran); Paul Anderson (SPREP); and Aylem Hernández Avila (National System of Protected Areas, Cuba).

Other Data. Additional data sets and assistance with spatial analysis were provided by the following people: Dan Russell (CleanCruising.com.au); Daniel Hesselink and Qyan Tabek (HotelbyMaps.com); Gregory Yetman (Center for International Earth Science Information Network); Siobhan Murray and Uwe Deichmann (World Bank); and Susan Minnemeyer, Florence Landsberg, Andrew Leach, and Lauriane Boisrobert (WRI). The global map of mangrove forests was developed by the partners of the *World Atlas of Mangroves*,[34] including the International Society for Mangrove Ecosystems, UNEP-WCMC, the Food and Agriculture Organization of the United Nations, and TNC. The International Tropical Timber Organization funded this work.

We would like to acknowledge all of those who contributed reef stories, who are named throughout the report and on the Reefs at Risk website with their story, including Enric Sala (National Geographic); Steven Victor (TNC-Palau); Annie Reisewitz and Jessica Carilli (Scripps, UCSD); Ronaldo Francini-Filho and Fabiano Thompson (Universidade Federal da Paraiba); Rodrigo Moura (CI-Brazil); Charles Sheppard (University of Warwick); Michael Gawel (Guam EPA); Sandrine Job (Independent); Sue Wells (Independent); Jason Vains and John Baldwin (Great Barrier Reef Marine Park Authority); Joanne Wilson and Purwanto (TNC); Wahyu Rudianto (Wakatobi National Park Authority); Veda Santiadji (WWF-Indonesia); Saharuddin Usmi (KOMUNTO, Wakatobi National Park); David Medio (Halcrow Group Ltd); Jamie Monty and Chantal Collier (Florida Dept. of Environmental Protection); Leona D'Agnes, Francis Magbanua, and Joan Castro (PATH Foundation Philippines); Stacy Jupiter (WCS-Fiji); Heidi Williams (Coral Reef Alliance); Susan Ruffo and Alison Green (TNC); James Maragos (USFWS); Ruben Torres (Reef Check DR); Jennie Mallela (ARC Centre for Coral Reef Studies and The Australia National University); Jorge Cortes (University of Costa Rica); Hector Guzman (Smithsonian Tropical Research Institute); Silvianita Timotius (The Indonesia Coral Reef Foundation); Idris, Estradivari, Mikael Prastowo and Muh. Syahrir (TERANGI); and Zaki Moustafa (Duke University).

We would like to thank the following formal reviewers of the report, who provided valuable comments on the manuscript and maps: Helen Fox (WWF), Ove Hoegh-Guldberg (University of Queensland), Liza Karina Agudelo (United Nations Foundation), Caroline Rogers (U.S. Geological Survey), and Jerker Tamelander (IUCN).

Internal reviewers from WRI include Maggie Barron, Mark Holmes, Hilary McMahon, Mindy Selman, Norbert Henninger, Heather McGray, and John Talberth. Special thanks to Craig Hanson, David Tomberlin, and Polly Ghazi for their many reviews of the draft and steady encouragement, and to Ashleigh Rich for her skillful management of the review process.

The following people reviewed specific parts of the text, reviewed regional maps, or provided general support: Tim McClanahan (WCS), Caroline Vieux (SPREP), James Maragos (US Fish and Wildlife Service), David Souter (Reef and Rainforest Research Centre), Judy Lang (Independent), David Medio (Halcrow Group Ltd), Annadel Cabanban (Sulu-Celebes/Sulawesi Seas Sustainable Fisheries Management Project), Abigail Moore and Samliok Ndobe (LP3L Talinti), Beatrice Padovani (Universidade Federal de Pernambuco), Sheila McKenna (Independent), Melanie McField (Smithsonian Institution), Marines Millet Encalada and Ricardo Gomez (National Marine Park of Cozumel), Pedro Alcolado (Oceanology Institute of Cuba), Nishanti Perera and Ramasamy Venkatesan (South Asian Seas Programme), Abigail Alling and Orla Doherty (Planetary Coral Reef Foundation), Dessy Anggraeni (Sustainable Fisheries Partnership), Yvonne Sadovy (University of Hong Kong), Laurie Raymundo (University of Guam), and Linwood Pendleton (Duke University).

In addition to many of those already mentioned, the following people provided valuable input through participation in one of the three *Reefs at Risk* threat analysis workshops. At the Washington, DC workshop: Barbara Best (USAID); Amie Brautigam (WCS); Andrew Bruckner (NOAA); Marea Hatziolos, Daniel Mira-Salama, Natsuko Toba, and Walter Vergara (World Bank); Will Heyman (Texas A&M); Charles Huang (WWF); Karen Koltes (US Dept. of Interior); Bruce Potter (Island Resources Foundation); Jean Wiener (Fondation pour la Protection de la Biodiversite Marine); Amanda Williams (Living Oceans Foundation); and Patricia Bradley, Dan Campbell, Bill Fisher, Suzanne Marcy, Leah Oliver, Debbie Santavay, Jordan West, and Susan Yee (EPA). At the Fiji workshop: Monifa Fiu (WWF-South Pacific), Louise Heaps (WWF-Fiji); Philippe Gerbeaux, Padma Narsey-Lal, and Kelvin Passfield (IUCN-Oceania); Stuart Gow (Fiji Islands Hotel and Tourism Assoc.); Naushad Yakub (WCS-Fiji); Jens Kruger (Pacific Islands Applied Geoscience Commission); Ed Lovell, Semisi Meo, Randy Thaman, Posa Skelton, and Joeli Veitayaki (USP); Franck Magron (SPC); Peter Ramohia (TNC); Chinnamma Reddy and Helen Sykes (Resort Support); Fatima Sauafea-Leau (NOAA); Ron Vave (LMMA Network); Caroline Vieux (SPREP); and Laurent Wantiez (University of New Caledonia). Special thanks to Cherie Morris, Robin South, and Shirleen Bala (USP) for coordinating and hosting the Fiji workshop. At the International Marine Conservation Congress workshop in Fairfax, VA: Hyacinth Armstrong (Bucco Reef Trust); Billy Causey and Susie Holst (NOAA); Eric Clua (SPC/CRISP); Richard Huber (Organization of American States); Esther Peters (George Mason University); Erica Rychwalski (TNC); Bernard Salvat and Francis Staub (ICRI); and Sean Southey (RARE Conservation).

We would also like to thank the following people who provided input and other helpful support: Andrew Baker (University of Miami); Nicola Barnard and Louisa Wood (UNEP-WCMC); David Sheppard (SPREP); Nadia Bood (WWF-Central America); Jon Brodie (James Cook University); Martin Callow (WCS-Fiji); Pip Cohen (ReefBase Pacific); Andy Hooten (AJH Environmental Services); Taholo Kami (IUCN Oceania); Suzanne Livingstone (Old Dominion University); Caleb McClennen (WCS); and Rashid Sumaila and Dirk Zeller (UBC).

Many other staff at WRI contributed to this project through publication, financial management, and outreach and assistance, including Beth Bahs-Ahern, Hyacinth Billings, Liz Cook, Laura Lee Dooley, Kathy Doucette, Tim Herzog, Robin Murphy, and Michael Oko. We appreciate the early and steady encouragement provided by Janet Ranganathan, Jonathan Lash, Dan Tunstall, and Manish Bapna.

The report was edited by Polly Ghazi (WRI) and Bob Livernash (independent). The report was embellished through the layout by Maggie Powell and the beautiful photographs provided by Wolcott Henry, Richard Ling, Stacy Jupiter, Steve Lindfield, Dave Burdick, Michael Emerson, Karen Koltes, Freda Paiva, Tewfik Alex, the Reef Check Foundation, GBRMPA, ARC Center of Excellence for Coral Reef Studies, Nguna Pela MPA Network, ReefBase, and many photographers using Marine Photobank, who are credited throughout this report.

PHOTO: RICHARD LING

Executive Summary

CORAL REEFS: VALUABLE BUT VULNERABLE

Coral reefs, the "rain forests of the sea," are among the most biologically rich and productive ecosystems on earth. They also provide valuable ecosystem benefits to millions of coastal people. They are important sources of food and income, serve as nurseries for commercial fish species, attract divers and snorkelers from around the world, generate the sand on tourist beaches, and protect shorelines from the ravages of storms.

However, coral reefs face a wide and intensifying array of threats—including impacts from overfishing, coastal development, agricultural runoff, and shipping. In addition, the global threat of climate change has begun to compound these more local threats to coral reefs in multiple ways. Warming seas have already caused widespread damage to reefs, with high temperatures driving a stress response called coral bleaching, where corals lose their colorful symbiotic algae, exposing their white skeletons. This is projected to intensify in coming decades. In addition, increasing carbon dioxide (CO_2) emissions are slowly causing the world's oceans to become more acidic. Ocean acidification reduces coral growth rates and, if unchecked, could reduce their ability to maintain their physical structure. With this combination of local threats plus global threats from warming and acidification, reefs are increasingly susceptible to disturbance or damage from storms, infestations, and diseases. Such degradation is typified by reduced areas of living coral, increased algal cover, reduced species diversity, and lower fish abundance.

Despite widespread recognition that coral reefs around the world are seriously threatened, information regarding which threats affect which reefs is limited, hampering conservation efforts. Researchers have studied only a small percentage of the world's reefs; an even smaller percentage have been monitored over time using consistent and rigorous methods. The World Resources Institute's *Reefs at Risk* series was initiated in 1998 to help fill this knowledge gap by developing an understanding of the location and spread of threats to coral reefs worldwide, as well as illustrating the links between human activities, human livelihoods, and coral reef ecosystems. With this knowledge, it becomes much easier to set an effective agenda for reef conservation.

PHOTO: MARK SPALDING

PURPOSE AND GOAL OF *REEFS AT RISK REVISITED*

Under the *Reefs at Risk Revisited* project, WRI and its partners have developed a new, detailed assessment of the status of and threats to the world's coral reefs. This information is intended to raise awareness about the location and severity of threats to coral reefs. These results can also catalyze opportunities for changes in policy and practice that could safeguard coral reefs and the benefits they provide to people for future generations.

Reefs at Risk Revisited is a high-resolution update of the original global analysis, *Reefs at Risk: A Map-Based Indicator of Threats to the World's Coral Reefs.*[1] *Reefs at Risk Revisited* uses a global map of coral reefs at 500-m resolution, which is 64 times more detailed than the 4-km resolution map used in the 1998 analysis, and benefits from improvements in many global data sets used to evaluate threats to reefs (most threat data are at 1 km resolution, which is 16 times more detailed than those used in the 1998 analysis). Like the original *Reefs at Risk*, this study evaluates threats to coral reefs from a wide range of human activities. For the first time, it also includes an assessment of climate-related threats to reefs. In addition, *Reefs at Risk Revisited* includes a global assessment of the vulnerability of nations and territories to coral reef degradation, based on their dependence on coral reefs and their capacity to adapt.

WRI led the *Reefs at Risk Revisited* analysis in collaboration with a broad partnership of more than 25 research, conservation, and educational organizations. Partners have provided data, offered guidance on the analytical approach, contributed to the report, and served as critical reviewers of the maps and findings.

The outputs of *Reefs at Risk Revisited* (report, maps, and spatial data sets) will be valuable to many users. Marine conservation practitioners, resource managers, policymakers and development agencies can use these tools to identify opportunities to protect reefs, set priorities, and plan interventions. Businesses that rely on or affect coral reef ecosystems can use this information to mitigate risks and protect their long-term economic interests. Educators can share this knowledge, thereby planting the seeds for a new generation of marine conservationists. The media can use it for its immediate and important news message, and as a basis for future research and communications. Overall, it is our hope that *Reefs at Risk Revisited* will clearly communicate what is at stake: why coral reefs are critically important and why it is essential that threats to reefs be reduced through better management practices and policies that protect these valuable ecosystems.

BOX ES-1. THREAT ANALYSIS METHOD

Human pressures on coral reefs are categorized throughout the report as either "local" or "global" in origin. These categories are used to distinguish between threats from human activities near reefs, which have a direct and relatively localized impact, versus threats that affect reefs indirectly, through human impacts on the global climate and ocean chemistry.

Local threats addressed in this analysis:

- Coastal development, including coastal engineering, land filling, runoff from coastal construction, sewage discharge, and impacts from unsustainable tourism.
- Watershed-based pollution, focusing on erosion and nutrient fertilizer runoff from agriculture delivered by rivers to coastal waters.
- Marine-based pollution and damage, including solid waste, nutrients, toxins from oil and gas installations and shipping, and physical damage from anchors and ship groundings.
- Overfishing and destructive fishing, including unsustainable harvesting of fish or invertebrates, and damaging fishing practices such as the use of explosives or poisons.

Global threats addressed in this analysis:

- Thermal stress, including warming sea temperatures, which can induce widespread or "mass" coral bleaching.
- Ocean acidification driven by increased CO_2 concentrations, which can reduce coral growth rates.

The four local threats to coral reefs were modeled separately, and subsequently combined in the *Reefs at Risk* integrated local threat index. The modeling approach is an extension and refinement of the one used in our previous analyses, and benefited from the input of more than 40 coral reef scientists and climate experts. For each local threat, a proxy indicator was developed by combining data reflecting "stressors," such as human population density and infrastructure features (including the location and size of cities, ports, and hotels), as well as more complex modeled estimates such as sediment input from rivers. For each stressor, distance-based rules were developed, where threat declines as distance from the stressor increases. Thresholds for low, medium, and high threats were developed using available information on observed impacts to coral reefs.

Local threats were modeled at WRI; data and models for global threats were obtained from external climate experts. Climate-related stressors are based on data from satellite observations of sea surface temperature, coral bleaching observations, and modeled estimates of future ocean warming and acidification. Input from coral reef scientists and climate change experts contributed to the selection of thresholds for the global threats.

Modeled outputs were further tested and calibrated against available information on coral reef condition and observed impacts on coral reefs. All threats were categorized as low, medium, or high, both to simplify the findings and to enable comparison between findings for different threats. In the presentation of findings, "threatened" refers to coral reefs classified at medium or high threat.

Full technical notes, including data sources and threat category thresholds, and a list of data contributors are available online at www.wri.org/reefs. Data sources are also listed in Appendix 2.

KEY FINDINGS

1. **The majority of the world's coral reefs are threatened by human activities.**
   - More than 60 percent of the world's reefs are under immediate and direct threat from one or more local sources —such as overfishing and destructive fishing, coastal development, watershed-based pollution, or marine-based pollution and damage.
   - Of local pressures on coral reefs, overfishing—including destructive fishing—is the most pervasive immediate threat, affecting more than 55 percent of the world's reefs. Coastal development and watershed-based pollution each threaten about 25 percent of reefs. Marine-based pollution and damage from ships is widely dispersed, threatening about 10 percent of reefs.
   - Approximately 75 percent of the world's coral reefs are rated as threatened when local threats are combined with thermal stress, which reflects the recent impacts of rising ocean temperatures, linked to the widespread weakening and mortality of corals due to mass coral bleaching (Figure ES-1, column 6).

2. **Local threats to coral reefs are the most severe in Southeast Asia and least severe in Australia (Figure ES-2).**
   - Of the six coral reef regions shown in Map ES-1, local pressure on coral reefs is highest in Southeast Asia, where nearly 95 percent of reefs are threatened, and about 50 percent are in the high or very high threat category. Indonesia, second only to Australia in the total area of coral reefs that lie within its jurisdiction, has the largest area of threatened reef, followed by the Philippines. Overfishing and destructive fishing pressure drive much of the threat in this region, followed by watershed-based pollution and coastal development.
   - In the Atlantic region, more than 75 percent of reefs are threatened, with more than 30 percent in the high or very high threat category. In more than 20 countries or territories in the region—including Florida (United States), Haiti, the Dominican Republic, and Jamaica—all reefs are rated as threatened. The Bahamas have the largest area of reef rated as low threat in this region. Overfishing is the most pervasive threat, but marine-based pollution and damage, coastal development, and watershed-based pollution also pose significant threats.
   - In the Indian Ocean, more than 65 percent of reefs are threatened by local activities, with nearly 35 percent under high or very high threat. The Maldives, the Chagos Archipelago, and the Seychelles have the largest area of reefs under low threat in the region. Overfishing is the most widespread threat, but land-based pollution and coastal development also elevate overall pressure.
   - In the seas of the Middle East, 65 percent of reefs are at risk from local threats, with more than 20 percent rated in the high or very high threat category. In Yemen, Qatar, Bahrain, Iran, Djibouti, and Kuwait, more than 95 percent of reefs are threatened. In this region, all four threats add significant pressure.
   - Although the wider Pacific region has long enjoyed relatively low pressure on coastal resources, almost 50 percent of reefs are currently considered threatened, with about 20 percent rated as high or very high. French Polynesia, the Federated States of Micronesia, Hawaii (United States), and the Marshall Islands have some of

FIGURE ES-1. REEFS AT RISK WORLDWIDE BY CATEGORY OF THREAT

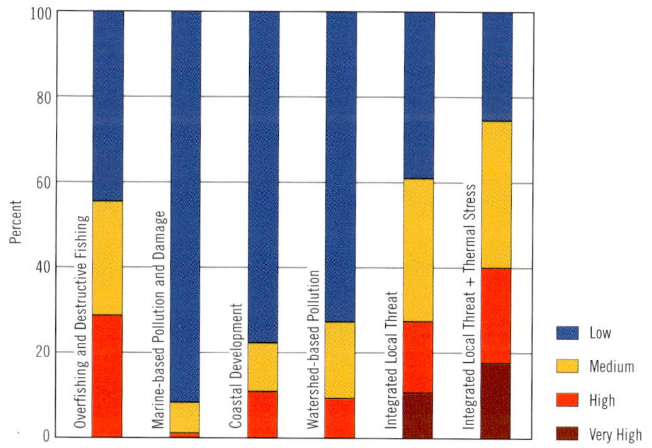

Notes: Individual local threats are categorized as low, medium, and high. These threats are integrated to reflect cumulative stress on reefs. Reefs with multiple high individual threat scores can reach the very high threat category, which only exists for integrated threats. The fifth column, integrated local threats, reflects the four local threats combined. The right-most column also includes thermal stress during the past ten years. This figure summarizes current threats; future warming and acidification are not included.

MAP ES-1. MAJOR CORAL REEF REGIONS OF THE WORLD AS DEFINED FOR THE *REEFS AT RISK REVISITED* ANALYSIS

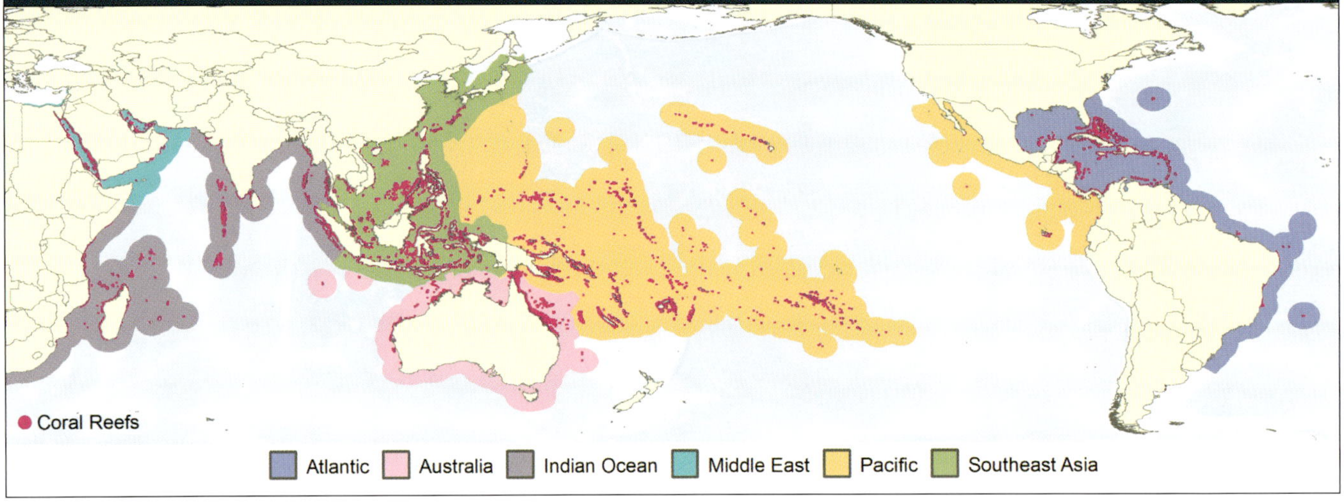

FIGURE ES-2. REEFS AT RISK FROM INTEGRATED LOCAL THREATS BY REGION

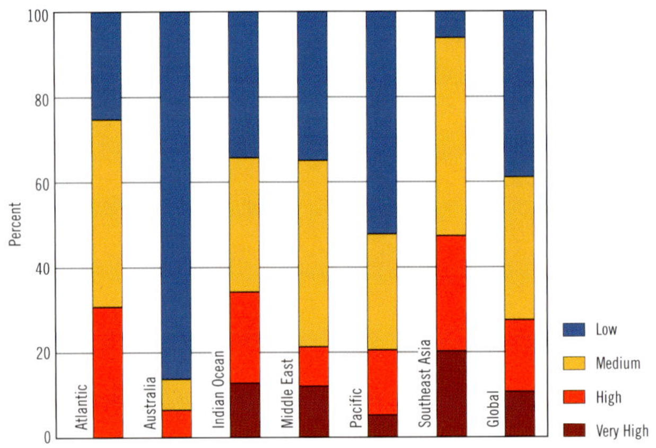

Note: Integrated local threats consist of the four local threats—overfishing and destructive fishing, marine pollution and damage, coastal development, and watershed-based pollution.

the lowest overall threat ratings (under 30 percent threatened.) Overfishing and runoff from land-based sources are the predominant threats, though coastal development is also a major pressure in some areas.

- Australia's reefs are the world's least threatened, with an estimated 14 percent threatened by local activities and just over 1 percent at high or very high threat. Our analysis identifies both marine-based pollution and watershed-based pollution as the dominant threats, but vast areas of reef are remote from such impacts.

3. **Threat levels have increased dramatically over a ten-year period.**

   - A separate analysis enabling a direct comparison of changes in threats over time shows that the percent of reefs rated as threatened has increased by 30 percent in the 10 years since the first *Reefs at Risk* analysis (comparing data from 1997 and 2007), with increases in all local threat categories and in all regions.
   - **By local threat:** The greatest driver of increased pressure on reefs since 1998 has been an 80 percent increase in the threat from overfishing and destructive fishing, most significantly in the Pacific and Indian Ocean regions. This change is largely due to the growth in coastal populations living near reefs. Pressure on reefs from coastal development, watershed-based pollution, and marine-based pollution and damage has also increased dramatically above 1998 levels.
   - **By region:** In the Pacific and Indian oceans, many reefs formerly classified as low threat are now threatened, largely reflecting increased overfishing pressure. In the Middle East, Southeast Asia, and the Atlantic over the past ten years, extensive areas of reefs have been pushed from medium threat into higher threat categories through a combination of local threats. Australia had the smallest increase in local pressure on reefs over the ten-year period.

4. **Changes in climate and in ocean chemistry represent significant and growing threats.**

 - **Impact of CO_2**: Rising concentrations of CO_2 and other greenhouse gases in the atmosphere have led to warming of the atmosphere and, as a result, an increase in sea surface temperatures. Mass coral bleaching, a stress response to warming waters, has occurred in every region and is becoming more frequent as higher temperatures recur. Extreme bleaching events kill corals outright, while less extreme events can weaken corals, affecting their reproductive potential, reducing growth and calcification, and leaving them vulnerable to disease. These effects also compound the local threats described above. Managing this threat is particularly challenging because it does not arise from local human actions, but from global changes to the atmosphere as a result of human activities.
 - **Thermal stress:** Our projections suggest that during the 2030s roughly half of reefs globally will experience thermal stress sufficient to induce severe bleaching in most years. During the 2050s, this percentage is expected to grow to more than 95 percent. These projections assume that greenhouse gas emissions continue on current trajectories and local threats are not addressed. Although coral reefs can recover from infrequent and mild bleaching, this degree of high, regular stress presents a significant risk of irreversible damage.
 - **Rising acidity**: Rising levels of CO_2 in the oceans are altering ocean chemistry and increasing the acidity of ocean water, reducing the saturation level of aragonite, a compound corals need to build their skeletons. By 2030, fewer than half the world's reefs are projected to be in areas where aragonite levels are ideal for coral

FIGURE ES-3. REEFS AT RISK: PRESENT, 2030, AND 2050

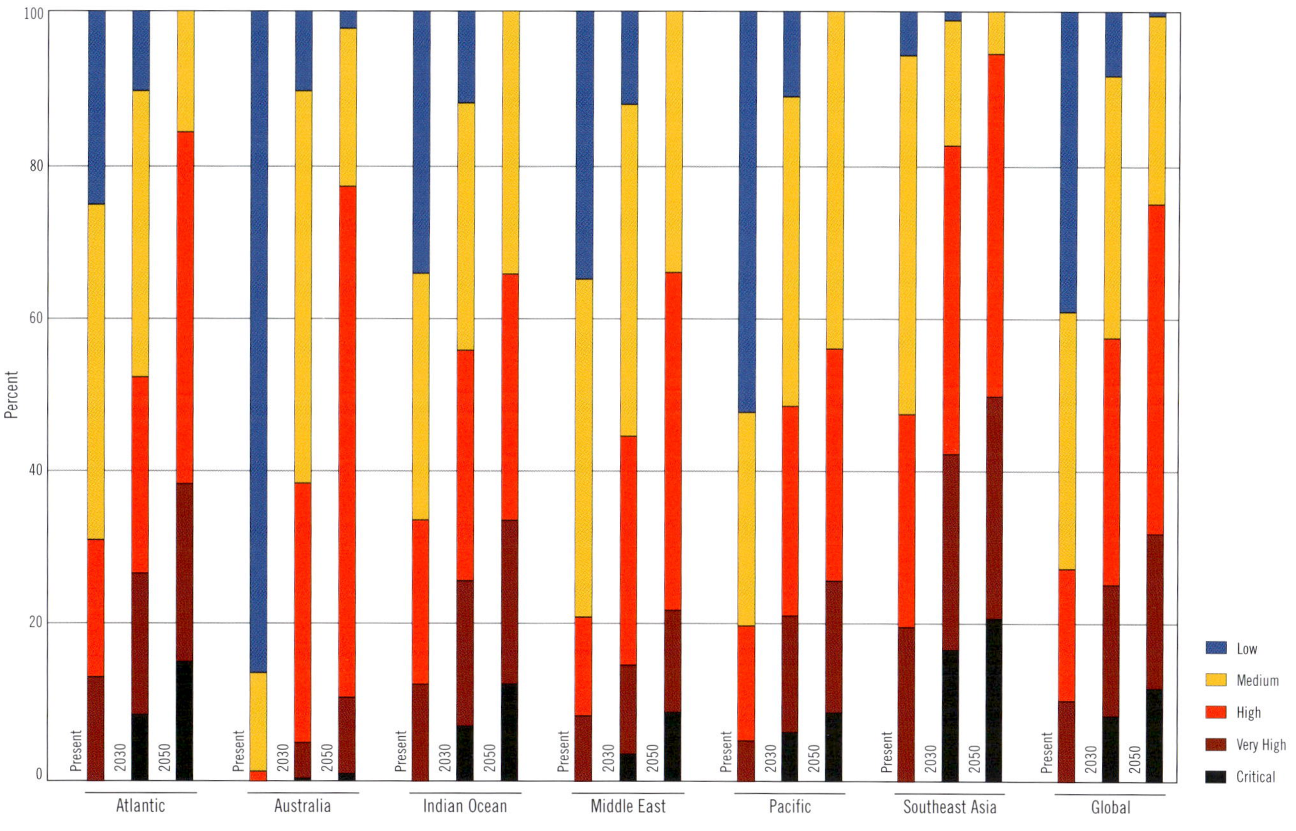

Note: "Present" represents the *Reefs at Risk* integrated local threat index, without past thermal stress considered. Estimated threats in 2030 and 2050 use the present local threat index as the base and also include projections of future thermal stress and ocean acidification. The 2030 and 2050 projections assume no increase in local pressure on reefs, and no reduction in local threats due to improved policies and management.

growth, suggesting that coral growth rates could be dramatically reduced. By 2050, only about 15 percent of reefs will be in areas where aragonite levels are adequate for coral growth.

- **Combined impacts**: The combined impacts of ocean warming and acidification will increase the threat levels on more than half of all reefs by 2030, pushing the percentage of threatened reefs to more than 90 percent by 2030. By 2050, nearly all reefs will be affected by warming and acidification and almost all reefs will be classified as threatened, assuming there is no change in local pressure on reefs (Figure ES-3).

5. **While over one quarter of the world's coral reefs are within protected areas, many are ineffective or only offer partial protection.**

- Approximately 27 percent of the world's reefs are located inside marine protected areas (MPAs). This coverage includes strictly controlled marine reserves, locally managed marine areas, and sites where management controls only one or two types of threat. Of the reef area inside MPAs, more than half is in Australia. Outside Australia, only 16 percent of coral reefs are within MPAs.
- We identified 2,679 MPAs in coral reef areas and were able to rate nearly half, including most of the larger sites, for their effectiveness in reducing the threat of overfishing. Of those rated, 15 percent of sites were rated as effective, 38 percent as partially effective, and 47 percent as ineffective.
- Based on these ratings, only 6 percent of the world's coral reefs are located in effectively managed MPAs and 73 percent are located outside MPAs (Figure ES-4). Increasing the MPA coverage and efficacy thus remains a priority for most areas.
- The coverage of MPAs is strongly biased away from areas of greatest threat, limiting their potential for reducing threats in areas of heavy human pressure.

FIGURE ES-4. CORAL REEFS BY MARINE PROTECTED AREA COVERAGE AND EFFECTIVENESS LEVEL

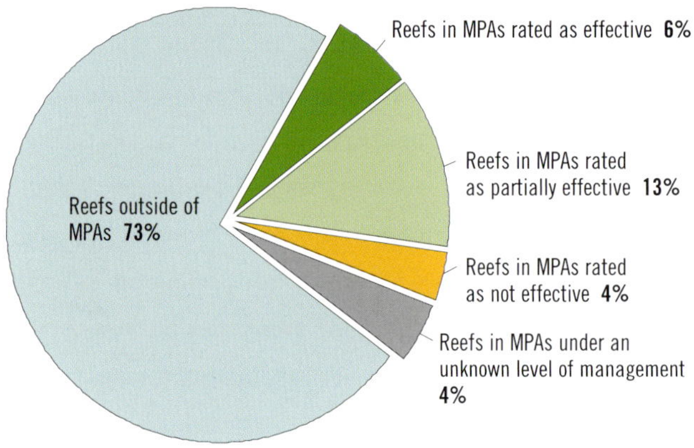

Note: The global area of coral reefs is 250,000 sq km (which represents 100% on this chart), of which 67,350 sq km (27%) is inside MPAs.

6. **Dependence on coral reefs is high in many countries, especially small-island nations.**

- Worldwide, approximately 850 million people live within 100 km of reefs, many of whom are likely to derive some benefits from the ecosystem services they provide. More than 275 million people reside in the direct vicinity of coral reefs (within 30 km of reefs and less than 10 km from the coast), where livelihoods are most likely to depend on reefs and related resources.
- Of 108 countries and territories studied, the most reef-dependent were almost all small-island states, many located in the Pacific and the Caribbean (Map ES-2).
- Populous Asian nations, such as Indonesia and the Philippines, account for the greatest absolute numbers of reef fishers. Relative to population size, many of the countries with high participation in reef fisheries are in the Pacific.
- At least 94 countries and territories benefit from reef tourism; in 23 of these, reef tourism accounts for more than 15 percent of gross domestic product (GDP).
- More than 150,000 km of shoreline in 100 countries and territories receive some protection from reefs, which reduce wave energy and associated erosion and storm damage.

MAP ES-2. SOCIAL AND ECONOMIC DEPENDENCE ON CORAL REEFS

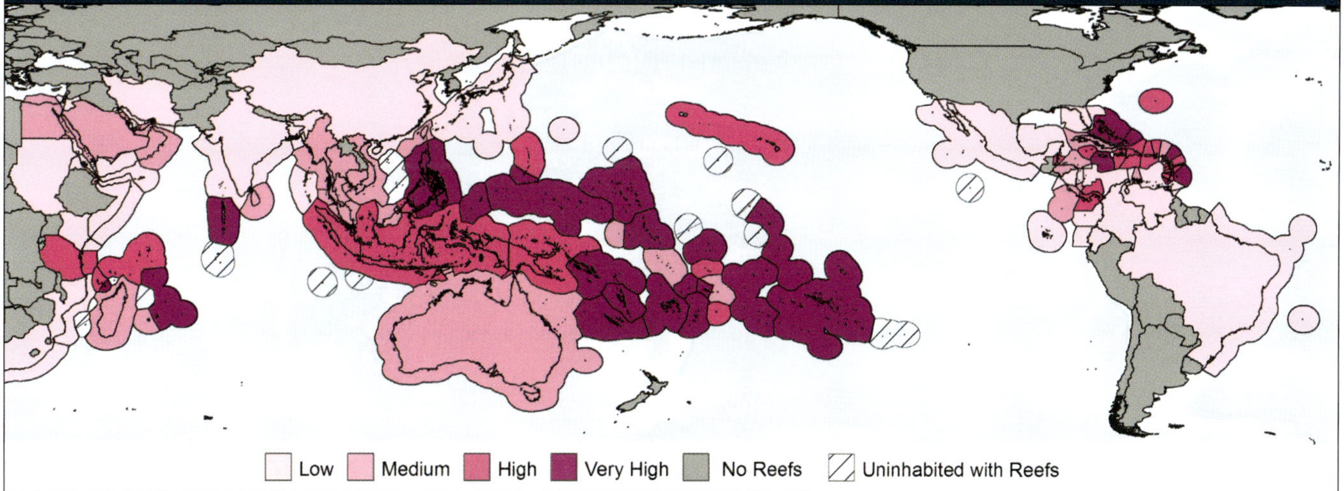

Note: Reef dependence is based on reef-associated population, reef fisheries employment, nutritional dependence on fish and seafood, reef-associated export value, reef tourism, and shoreline protection from reefs. Countries and territories are categorized according to quartiles.

7. **Degradation and loss of reefs will result in significant social and economic impacts.** Vulnerability to reef loss was assessed for 108 inhabited reef countries and territories, based on exposure to reef threats, dependence on ecosystem services (food, livelihoods, exports, tourism, and shoreline protection), and adaptive capacity (ability to cope with the effects of degradation).
   - The 27 countries and territories identified as highly vulnerable to reef loss are spread across the world's reef regions (Figure ES-5). Nineteen are small-island states.
   - Nine countries—Haiti, Grenada, the Philippines, Comoros, Vanuatu, Tanzania, Kiribati, Fiji, and Indonesia—are most vulnerable to the effects of coral reef degradation. They have high ratings for exposure to reef threat and reef dependence, combined with low ratings for adaptive capacity. These countries merit the highest priority for concerted development efforts to reduce reliance on reefs and to build adaptive capacity, alongside reducing immediate threats to reefs.

CONCLUSIONS AND RECOMMENDATIONS

This report presents a deeply troubling picture of the world's coral reefs. Local human activities already threaten the majority of reefs in most regions, and the accelerating impacts of global climate change are compounding these problems. The extent and severity of threats to reefs, in combination with the critically important ecosystem services they provide, point to an urgent need for action. The report offers reason for hope: reefs around the world have shown a capacity to rebound from even extreme damage, while active management is protecting reefs and aiding recovery in some areas.

However, we need to improve, quickly and comprehensively, on existing efforts to protect reefs and the services they provide humanity. It is encouraging that our collective ability to do so has become stronger, with new management tools, increased public understanding, better communications, and more active local engagement. We hope this new report will spur further action to save these critical ecosystems. The array of measures to deal with the many threats to reefs must be comprehensive. Local threats must be tackled head-on with direct management interventions, while efforts to quickly and significantly reduce greenhouse gas emissions are of paramount concern not only for reefs, but for nature and humanity as a whole. At the same time, we may be able to "buy time" for coral reefs in the face of cli-

FIGURE ES-5. DRIVERS OF VULNERABILITY IN HIGHLY VULNERABLE NATIONS AND TERRITORIES

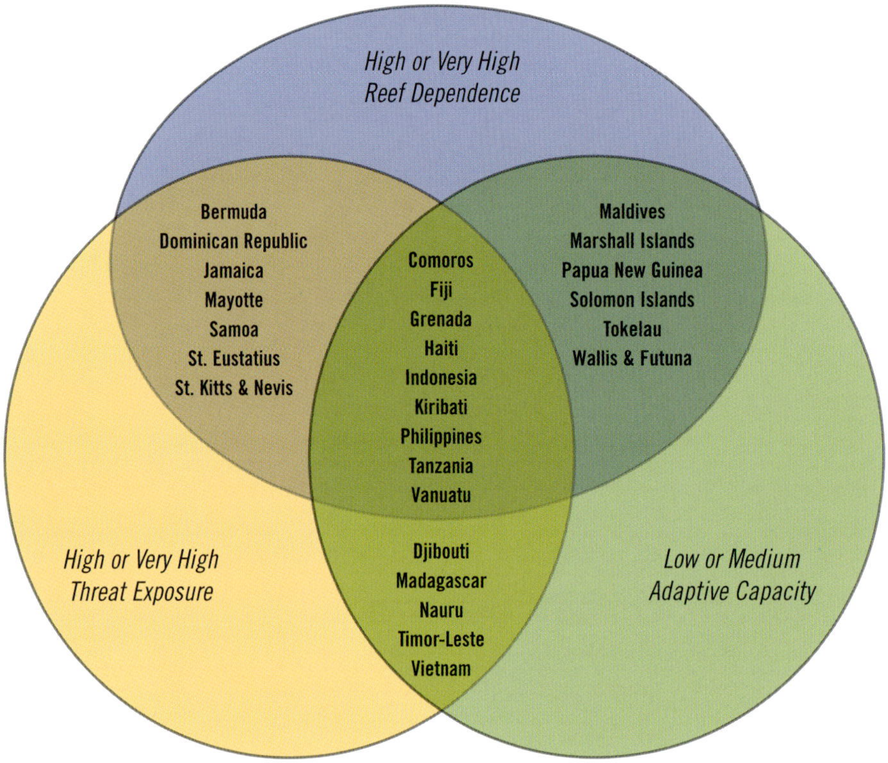

Note: Countries or territories within the yellow circle are highly or very highly exposed to reef threat; those within the blue circle are highly or very highly reef-dependent; and those within the green circle have low or medium adaptive capacity. Only the 27 very highly vulnerable countries and territories are shown.

mate change, through local-scale measures to increase reef resilience to climate-related threats.

Toward these aims, we recommend the following specific actions involving a broad range of stakeholders at the local, national, regional, and international scales:

- Mitigate threats from local human activities.
    - ***Reduce unsustainable fishing*** by addressing the underlying social and economic drivers of overfishing; establishing sustainable fisheries management policies and practices; reducing excess fishing capacity and removing perverse subsidies; enforcing fishing regulations; halting destructive fishing; improving and expanding MPAs to maximize benefits; and involving stakeholders in resource management.
    - ***Manage coastal development*** through coastal zone planning and enforcement to prevent unsound land development; protecting coastal vegetation; implementing erosion-control measures during construction; improving sewage treatment; linking marine and terrestrial protected areas; and developing tourism in sustainable ways.
    - ***Reduce watershed-based pollution*** by reducing sediment and nutrient delivery to coastal waters through improved agriculture, livestock, and mining practices; minimizing industrial and urban runoff; and protecting and restoring riparian vegetation.
    - ***Reduce marine-based pollution and damage*** by reducing at-sea disposal of waste from vessels; increasing regulation of ballast discharge from ships; designating safe shipping lanes and boating areas; managing offshore oil and gas activities; and using MPAs to protect reefs and adjacent waters.

- **Manage for climate change locally.** A growing body of evidence has shown that by reducing local threats (including overfishing, nutrients, and sediment pollution), reefs may be able to recover more quickly from coral bleaching. Strategic planning to enhance local-scale reef resilience should target critical areas, building networks of protected areas that include (and replicate) different parts of the reef system, as well as include areas critical for future reef replenishment. Such efforts may represent an opportunity to "buy time" for reefs, until global greenhouse gas emissions can be curbed.

- **Develop integrated management efforts at ecosystem scales.** Plans that are agreed to by all sectors and stakeholders and that consider ecological relationships are most likely to avoid waste, repetition, and potential conflicts with other interventions and maximize potential benefits. For reefs, relevant approaches include ecosystem-based management, integrated coastal management, ocean zoning, and watershed management.

- **Scale up efforts through international collaboration**. At all scales, we need political will and economic commitment to reduce local pressures on reefs and promote reef resilience in the face of a changing climate. It is also critical to replicate successful local and national approaches, and work internationally, using tools such as transboundary collaboration and regional agreements, improved international regulations to govern trade in reef products, and international agreements such as the UN Convention on the Law of the Sea, which helps regulate fishing, and MARPOL, which controls pollution from ships.

- **Support climate change efforts**. Reef scientists recommend not only a stabilization of CO_2 and other greenhouse gas concentrations, but also a slight reduction from our current level of 388 ppm (2010) to 350 ppm, if large-scale degradation of reefs is to be avoided. Attaining this challenging target will take time, and require immense global efforts. There is a role to be played by all—individuals and civil society, NGOs, scientists, engineers, economists, businesses, national governments, and the international community—to address this enormous and unprecedented global threat.

- **Build consensus and capacity.** Closing the gap between knowledge and results depends on action within the following key areas:
  - ***Scientific research*** to build understanding of how particular reefs are affected by local activities and climate change and how different stressors may act in combination to affect reef species; to explore factors that confer resilience to reef systems and species; to assess the extent of human dependence on specific reef ecosystem services; and to determine the potential for coastal communities to adapt to expected change.
  - ***Education and communication*** to inform communities, government agencies, donors, and the general public about how current activities threaten reefs and why action is needed to save them, and to highlight examples of replicable conservation success.
  - ***Policy support*** to aid decisionmakers and planners in making long-term decisions that will affect the survival of coral reefs, as well as enhancing the ability of coastal communities to adapt to environmental changes and reef degradation.
  - ***Economic valuation*** to highlight the value of reefs and the losses associated with reef degradation, and to aid in assessing the longer-term costs and benefits of particular management and development plans.

PHOTO: STACY JUPITER

- ***Training and capacity building*** of reef stakeholders, to manage and protect reefs, understand and argue for their value, spread awareness, and reduce vulnerability in reef-dependent regions.
- ***Involvement of local stakeholders*** in the decision-making and management of reef resources.

■ **Individual action.** Regardless of whether you live near or far from a coral reef, you can take action to help coral reefs:

- ***If you live near coral reefs:***
  - Follow local laws and regulations designed to protect reefs and reef species.
  - If you fish, do it sustainably, avoiding rare species, juveniles, breeding animals, and spawning aggregations.
  - Avoid causing physical damage to reefs with boat anchors, or by trampling or touching reefs.
  - Minimize your indirect impacts on reefs by choosing sustainably caught seafood and reducing household waste and pollution that reaches the marine environment.
  - Help improve reef protection by working with others in your area to establish stronger conservation measures, participating in consultation processes for planned coastal or watershed development projects, and supporting local organizations that take care of reefs.
  - Tell your political representatives why protecting coral reefs is important.
- ***If you visit coral reefs:***
  - Choose sustainably managed, eco-conscious tourism providers.
  - Dive and snorkel carefully, to avoid physically damaging reefs.
  - Tell people if you see them doing something harmful to reefs.
  - Visit and make contributions to MPAs to support management efforts.
  - Avoid buying souvenirs made from corals and other marine species.
- ***Wherever you are***:
  - Choose sustainably caught seafood.
  - Avoid buying marine species that are threatened or may have been caught or farmed unsustainably.
  - Help to prioritize coral reefs, the environment, and climate change issues within your government
  - Support NGOs that conserve coral reefs and encourage sustainable development in reef regions.
  - Educate through example, showing your family, friends, and peers why reefs are important to you.
  - Reduce your carbon footprint.

PHOTO: KAREN KOLTES

Chapter 1. INTRODUCTION: ABOUT REEFS AND RISK

"One way to open your eyes is to ask yourself, 'What if I had never seen this before? What if I knew I would never see it again?' " – *Rachel Carson*

Coral reefs are one of the most productive and biologically rich ecosystems on earth. They extend over about 250,000 sq km of the ocean—less than one-tenth of one percent of the marine environment—yet they may be home to 25 percent of all known marine species.[2] About 4,000 coral reef-associated fish species and 800 species of reef-building corals have been described to date,[3] though these numbers are dwarfed by the great diversity of other marine species associated with coral reefs, including sponges, urchins, crustaceans, mollusks, and many more (Box 1.1).

Coral reefs exist within a narrow belt across the world's tropical oceans, where local conditions—climate, marine chemistry, ocean currents, and biology—combine to meet the exacting requirements of reef-building corals. A coral reef is both a physical structure and a diverse ecosystem. The physical structure is built up from the sea bed over centuries or millennia through the accumulated deposition of limestone-like (calcium carbonate) skeletons, laid down by reef-building corals. This structure, with a living veneer of corals on its surface, provides the basis for the incredible diversity of plant and animal species that live in and around it. Together, they form the coral reef ecosystem.

WHY REEFS MATTER

Dynamic and highly productive, coral reefs are not only a critical habitat for numerous species, but also provide essential ecosystem services upon which millions of people depend.

Food and livelihoods: One-eighth of the world's population—roughly 850 million people—live within 100 km of a coral reef and are likely to derive some benefits from the ecosystem services that coral reefs provide (Figure 1.1). More than 275 million people live very close to reefs (less than 10 km from the coast and within 30 km of reefs.)[4] Many of these people live in developing countries and island nations where dependence on coral reefs for food and livelihoods is high. Reef-associated fish species are an important source of protein, contributing about one-quarter of the total fish catch on average in developing countries.[5] A healthy, well-managed reef can yield between 5 and 15 tons of fish and seafood per square kilometer per year. [6, 7]

Tourism: Coral reefs are vital to tourism interests in many tropical countries, attracting divers, snorkelers, and recreational fishers. Reefs also provide much of the white sand for beaches. More than 100 countries and territories benefit from tourism associated with coral reefs, and tourism contributes more than 30 percent of export earnings in more than 20 of these countries.[8, 9]

Treatments for disease: Many reef-dwelling species have developed complex chemical compounds, such as venoms and chemical defenses, to aid their survival in these highly competitive habitats. Many such compounds harbor the potential for forming the basis of life-saving pharmaceuticals. Explorations into the medical application of reef-related compounds to date include treatments for cancer, HIV, malaria, and other diseases.[10] For example, scientists have synthesized an anti-cancer agent discovered in Caribbean sea squirts into a treatment for ovarian and other cancers.[11] Since only a small portion of reef life has been sampled, the potential for new pharmaceutically valuable discoveries is vast.[10]

Shoreline protection: Beyond their biological value, the physical structures of coral reefs protect an estimated 150,000 km of shoreline in more than 100 countries and territories.[12] Reefs dissipate wave energy, reducing routine erosion and lessening inundation and wave damage during storms. This function protects human settlements, infrastructure, and valuable coastal ecosystems such as seagrass meadows and mangrove forests.[13] Some countries—especially low-lying atolls such as the Maldives, Kiribati, Tuvalu, and the Marshall Islands—have been built entirely by coral reefs and would not exist but for their protective fringe.

LOCAL AND GLOBAL THREATS TO REEFS

Despite their importance, coral reefs face unprecedented threats throughout most of their range. Many reefs are already degraded and unable to provide the vital services on which so many people depend. Some threats are highly visible and occur directly on reefs. Levels of fishing are currently unsustainable on a large proportion of the world's reefs,[7, 16] and have led to localized extinctions of certain fish species, collapses and closures of fisheries, and marked ecological changes.[17, 18, 19] Many other threats are the result of human activities that occur far removed from the reefs. Forest clearing, crop cultivation, intensive livestock farming, and poorly planned coastal development have contributed increased sediments and nutrients to coastal waters, smothering some corals and contributing to overgrowth by algae. Pollution and waste from ships and from oil and gas exploitation further exacerbate the situation.

Beyond these extensive and damaging local-scale impacts, reefs are increasingly at risk from the global threats associated with rising concentrations of greenhouse gases in the atmosphere. Even in areas where local stresses on reefs are relatively minimal, warming seas have caused widespread damage to reefs through mass coral bleaching, which occurs when corals become stressed and lose, *en masse*, the zooxanthellae that live within their tissues and normally provide their vibrant colors. Increasing concentrations of carbon dioxide (CO_2) in the atmosphere, the result of deforestation and the burning of fossil fuels, are also changing the chemistry of the ocean surface waters. About one-third of the excess CO_2 in the atmosphere dissolves in the sea and, in so doing, causes ocean acidification—a decrease in the pH of seawater. This in turn creates changes to other seawater compounds, notably a reduction in the concentrations of carbonate ions, which are necessary for coral growth. If unchecked, this process will slow and then halt reef growth, and could cause coral skeletons and reefs to dissolve.[20]

It is rare for any reef to suffer only a single threat. More often the threats are compounded. For instance, overfishing eliminates a key herbivore, while runoff from agriculture supplies nutrients that cause a bloom in macroalgae, reducing the abundance or impairing the growth of coral and ultimately reducing the competitive ability of coral communities. A reef left vulnerable by one threat can be pushed to ecological collapse by the addition of a second.[21, 22]

AN URGENT PRIORITY: FILLING THE INFORMATION GAP

Despite widespread recognition that reefs are severely threatened, information regarding particular threats to particular reefs is limited. Only a fraction of reefs have been studied, and an even smaller percentage has been monitored over time using consistent methods. In a few areas—such as Jamaica, Florida, and Australia's Great Barrier Reef—changes in coral condition are well-documented. In most places, however, the availability of detailed information is limited, inhibiting effective management.

WRI initiated the *Reefs at Risk* series in 1998 to help fill this gap in knowledge and to link human activities on

BOX 1.1. CORAL REEF ECOSYSTEMS

The approximately 800 species of reef-building corals that inhabit tropical oceans are simple organisms. Individual coral animals known as polyps live in compact colonies of many identical individuals and secrete calcium carbonate to form a hard skeleton. Corals produce colonies in a multitude of shapes—huge boulders, fine branches, tall pillars, leafy clusters—and vibrant colors. These colonies build around and on top of one another, while sand and rubble fill the empty spaces. Calcareous algae also contribute by "gluing" together the matrix to form a solid three-dimensional structure. Thus, a coral reef is born.

A coral polyp has a simple tubular body with a ring of stinging tentacles around a central mouth. The polyps contain microscopic plants or algae (known as zooxanthellae) which live within their tissues. Corals filter food from the water using their tentacles, but they also rely heavily on their zooxanthellae, which use the sun's energy to synthesize sugars. The algae provide a critical source of food to the corals, enabling them to grow where nutrients are scarce, but restricting them to shallow waters, typically 50 meters or less in depth. Some coral species do not have zooxanthellae, and can thrive even in dark, cold or murky waters. In most places they do not build large structures, but coldwater coral reefs have recently been discovered in many areas of deep cold oceans. Unlike tropical coral reefs, these reefs have much lower diversity and are quite different ecosystems. (Coldwater reefs are not included in this analysis.)

Reef types: Scientists often describe reefs by the shape of the structures they build. Fringing reefs follow the coastline, tracing the shore tens or hundreds of meters from the coast. Barrier reefs lie far offshore, separated from the coast by wide, deep lagoons. Far out in the open ocean, some coral reefs mark the remains of what were once high islands, where atolls have been formed by the continued upward growth of corals, even as their original bedrock – ancient marine volcanoes—has sunk to form a lagoon.

Reef species: Living among the towers, canyons, and recesses of a typical coral reef structure are thousands of species of fish and invertebrates. Soft corals, whip corals, and sea fans are close relatives of the reef-builders, but lack their hard skeletons. Sea urchins and sea cucumbers are among the grazers. Sponges take a multitude of forms and, as immobile creatures, constantly filter the water for food. Over 4,000 fish species inhabit coral reefs.[14] Some butterflyfish and parrotfish feed on the corals themselves, while damselfish and others feed on plant life, and groupers prey on smaller fish and reef-dwellers. Crabs and lobsters are among the many nocturnal feeders. Worms and mollusks burrow inside the corals and rocks, collecting microscopic plankton.

Reef area: Coral reefs are found in the shallow seas of the tropics and subtropics. The total area that coral reefs inhabit globally is approximately 250,000 sq km,[15] which is roughly equivalent to the size of the United Kingdom.

FIGURE 1.1. NUMBER OF PEOPLE LIVING NEAR CORAL REEFS IN 2007

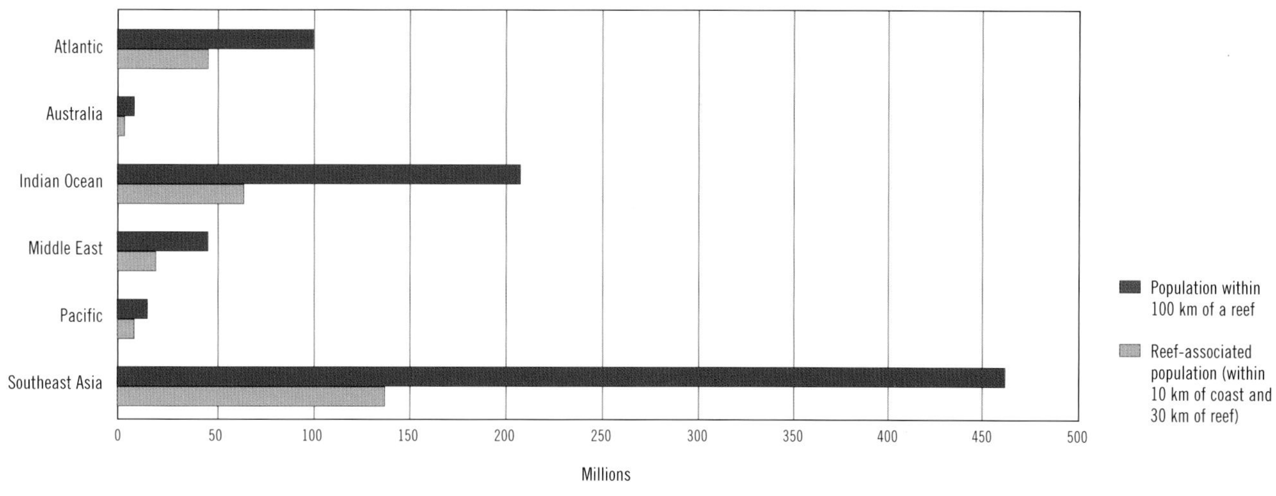

Source: WRI, using Landscan 2007 population data.

land and in the sea with their often unintended consequences. The 1998 report[1] came at a time when concerns about the declining status of coral reefs were already high. The work offered a first-ever quantification of those concerns, confirming that reefs were indeed threatened, and not only in areas where threats were well-known, but around the globe.

Since that first publication, considerable changes have taken place. Governments and the conservation community have increased efforts to protect and better manage reefs. Large numbers of marine protected areas (MPAs) have been established, fisheries management has improved in many areas, and reef-related concerns have been incorporated into coastal planning and watershed management. Unfortunately, in parallel, the drivers of threats to reefs have also continued to escalate—including growing populations, rising consumption, expanding agriculture, increasing trade and tourism, and accelerating greenhouse gas emissions.

Thirteen years after we released the first *Reefs at Risk*, our maps show clearly that the growth in threats has largely outpaced efforts to address those threats. Meanwhile, new threats have emerged, and others have expanded to new places, as forest clearing, agricultural expansion and intensification, population growth, and consumption have shifted and increased. In the mid-1990s, climate change was still perceived as a somewhat distant threat. However, in 1998, a powerful El Niño event[a] further increased sea surface temperatures that were already rising due to climate change, triggering the most severe and expansive coral bleaching event on record. Other mass-scale bleaching events have followed, and it appears that coral reefs are among the most sensitive of all major ecosystems on Earth to climate change. At the same time, we now have a greater understanding of the considerable dependence that many people and reef nations have on coral reefs for food security, employment, and income. An update of the global analysis is clearly necessary to identify and understand the effects and implications of changes to the world's reefs and to help guide targeted interventions.

a. El Niño events are cyclic oscillations of the ocean-atmosphere system in the tropical Pacific, resulting in unusually warm temperatures in the equatorial Pacific Ocean. These events can influence weather patterns around the world.

ABOUT THIS REPORT

This report is designed to support policymaking, management, coastal planning, and conservation efforts by providing the best and most detailed mapping of human pressure on coral reefs, using the most recent data available (most data sets date from 2007 to 2009).

It provides a summary of the analysis; additional materials and data sets are available online at www.wri.org/reefs. For the first time, this analysis explicitly includes threats from climate change and ocean acidification in the modeling, as well as an evaluation of how human pressure has changed over ten years (1998 to 2007). As the loss of associated goods and services and implications for reef-dependent people are central to our interests, the report also includes an analysis of the social and economic vulnerability of coral reef nations to reef degradation and loss. Specifically:

- Chapter 2 outlines the modeling methods and its limitations.
- Chapter 3 presents an overview of threats to the world's coral reefs, structured around six categories of threat.
- Chapter 4 summarizes the results of the global modeling of threats, presents an analysis of change in human pressure on reefs over the past ten years, and examines the implications of climate change and ocean acidification for coral reefs to 2050.
- Chapter 5 provides more detailed regional descriptions of the findings, and places these into a wider discussion of threats, status, and management for six major reef regions.
- Chapter 6 examines social and economic vulnerability of coral reef nations, with an emphasis on reef dependence, and a consideration of the economic values of coral reefs.
- Chapter 7 discusses management of coral reefs, including a review of the extent and effectiveness of tropical marine protected areas (MPAs).
- Chapter 8 provides conclusions and recommendations for actions needed at all levels to minimize threats and to halt global declines in coral reefs.

Chapter 2. PROJECT APPROACH AND METHODOLOGY

To quantify threats and to map where reefs are at greatest risk of degradation or loss, we incorporated more than 50 data sources into the analysis—including data on bathymetry (ocean depth), land cover, population distribution and growth rate, observations of coral bleaching, and location of human infrastructure. These data were consolidated within a geographic information system (GIS), and then used to model several broad categories of threat from human activities, climate change, and ocean acidification. In the absence of complete global information on reef condition, this analysis represents a pragmatic hybrid of monitoring observations and modeled predictions of reef condition.

Human pressures on coral reefs are categorized throughout the report as either "local" or "global" in origin. These categories are used to distinguish between threats that involve human activities near reefs that have a direct and relatively localized impact, versus threats that affect the reef environment indirectly through the cumulative impact of human activities on the global climate and ocean chemistry. Local threats addressed in this analysis are:

- Coastal development
- Watershed-based pollution
- Marine-based pollution and damage
- Overfishing and destructive fishing.

Global-level threats addressed are:

- Thermal stress (warming sea temperatures, which can induce coral bleaching)
- Ocean acidification (driven by increased CO_2, which can reduce coral growth rates).

This is the first *Reefs at Risk* project to incorporate data on these global-level threats. These data allow us not only to estimate current and imminent reef condition, but also to project trends well into the future. For the global-level threats, we did not develop new models, but rather incorporated existing data from partner organizations on past thermal stress, future thermal stress, and ocean acidification (Appendix 2). These data have enabled us to consider impacts to date and the potential future effects of ocean warming and acidification on reefs to 2030 and 2050 using climate projection scenarios.

The *Reefs at Risk Revisited* project delivers results as maps showing the distribution of local- and global-level threats to coral reefs. These threats are also consolidated into a single integrated index, which represents their combined impact on mapped reef locations. The analysis draws on a newly compiled global reef map—the most comprehensive and detailed rendition of global coral reef locations created

MAP 2.1. MAJOR CORAL REEF REGIONS OF THE WORLD AS DEFINED BY THE *REEFS AT RISK REVISITED* ANALYSIS

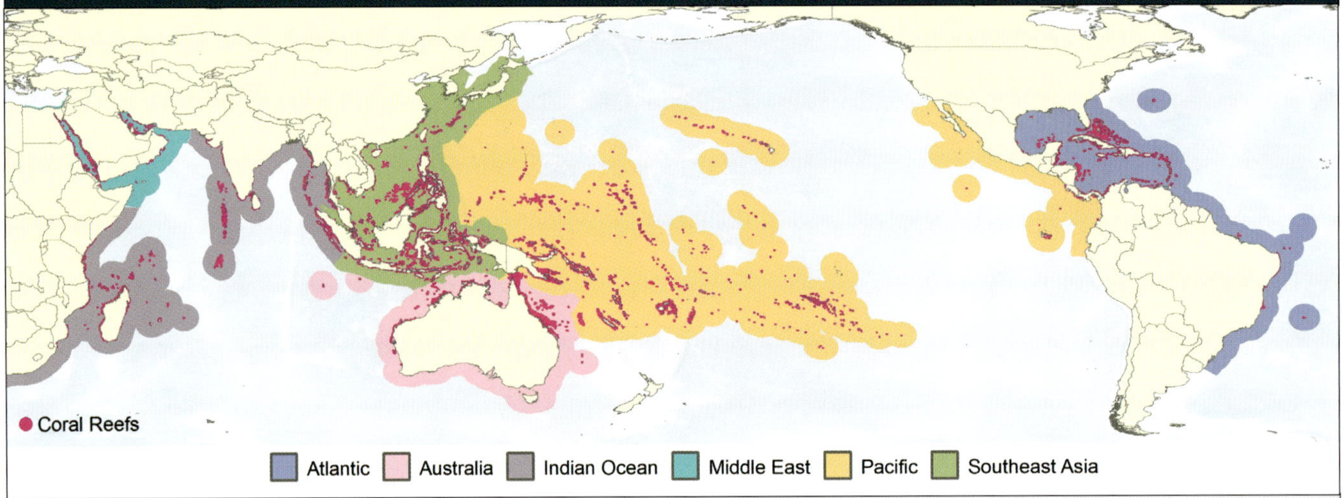

Source: WRI 2011.

FIGURE 2.1. DISTRIBUTION OF CORAL REEFS BY REGION

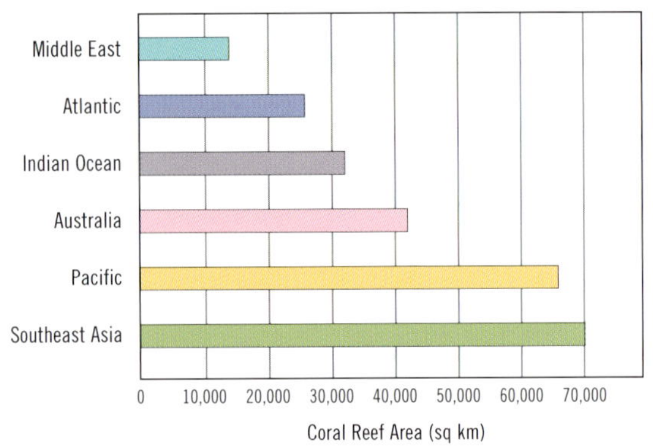

Note: Area of coral reefs (sq km) for each coral reef region of the world. The regions are shown in Map 2.1.
Sources: IMaRS/USF, IRD, NASA, UNEP-WCMC, WorldFish Center, WRI 2011.

to date—which we compiled into a 500-m resolution grid for modeling. Alongside mapped results, summary findings are presented for each of six major coral reef regions (Map 2.1).

Through the individual threat indicators and the integrated local threat index, *Reefs at Risk Revisited* estimates the level of human pressure on coral reefs. The index is not a direct measure of reef status or condition; some areas rated as threatened may have already suffered considerable loss or degradation, while others are still healthy. For healthy reefs, a high threat score is a measure of risk, a pointer to likely, even imminent, damage.[23] More typically, however, reefs that are threatened are already showing signs of damage—such as reduced live coral cover, increased algal cover, or reduced species diversity. Even in this case, it is important to realize that reef degradation is not a simple, step-wise change, but rather a cascade of ongoing changes. Even where degradation is already apparent, the models provide a critical reminder that future change will often make matters worse.

THREAT ANALYSIS METHOD

The four local threats to coral reefs were modeled separately, and later combined in the *Reefs at Risk* integrated local threat index. The modeling approach is an extension and refinement of that used in previous *Reefs at Risk* analyses, and benefited from input from more than 40 coral reef scientists and other experts. For each local threat, sources of stress that could be mapped were identified and combined into a proxy indicator that reflected the degree of threat. These "stressors" include human population density and infrastructure features such as location and size of cities, ports, and hotels, as well as more complex modeled estimates such as sediment inputs from rivers. For each stressor, distance-based rules were developed, such that threat declines as distance from the stressor increases. Thresholds for low, medium, and high threats were developed using available information on observed impacts to coral reefs. Table 2.1 provides a summary of the approach and limitations for modeling each local threat.

TABLE 2.1 REEFS AT RISK REVISITED ANALYSIS METHOD—PRESENT THREATS

Threat	Analysis Approach	Limitations
Coastal development	■ The threat to coral reefs from coastal development was modeled based on size of cities, ports, and airports; size and density of hotels; and coastal population pressure (a combination of population density, growth, and tourism growth).	■ Provides a good indicator of relative threat, but is likely to miss some (especially new) tourism locations. ■ Does not directly capture sewage discharge, but relies on population as a proxy for this threat.
Watershed-based pollution	■ The threat to reefs from land-based pollutants was modeled for over 300,000 watersheds (catchments) discharging to coastal waters. ■ Relative erosion rates were estimated across the landscape based on slope, land cover type, precipitation, and soil type. ■ Sediment delivery at the river mouth was estimated based on total erosion in the watershed, adjusted for the sediment delivery ratio (based on watershed size) and sediment trapping by dams and mangroves. ■ Sediment plume dispersion was modeled using a linear decay rate from the river mouth and was calibrated against actual sediment plumes observed from satellite data.	■ The model represents a proxy for sediment, nutrient, and pollutant delivery. ■ Nutrient delivery to coastal waters is probably underestimated due to a lack of spatial data on crop cultivation and fertilizer application. However, agricultural land is treated as a separate category of land cover, weighted for a higher influence. ■ The model does not incorporate nutrient and pollutant inputs from industry, or from intensive livestock farming, which can be considerable.
Marine-based pollution and damage	■ The indicator of threat from marine-based pollution and damage was based on the size and volume of commercial shipping ports, size and volume of cruise ship ports, intensity of shipping traffic, and the location of oil infrastructure.	■ Threat associated with shipping intensity may be underestimated because the data source is based on voluntary ship tracking, and does not include fishing vessels. ■ The threat model does not account for marine debris (such as plastics), discarded fishing gear, recreational vessels or shipwrecks, due to a lack of global spatial data on these threats.
Overfishing and destructive fishing	■ Threats to coral reefs from overfishing were evaluated based on coastal population density and extent of fishing areas (reef and shallow shelf areas), with adjustments to account for the increased demand due to proximity to large populations and market centers. Areas where destructive fishing occurs (with explosives or poisons) were also included, based on observations from monitoring and mapping provided by experts. ■ The threat estimate was reduced inside marine protected areas that had been rated by experts as having "effective" or "partially effective" management (meaning that a level of management is present that helps to guard ecological integrity).	■ Accurate, spatially referenced global data on fishing methods, catches, and number of fishers are not available; therefore, population pressure is used as a proxy for overfishing. ■ The model fails to capture the targeting of very high value species, which affects most reefs globally, but has fewer ecosystem impacts than wider scale overfishing. ■ Management effectiveness scores were only available for about 83% of the reefs within marine protected areas.
The four local threats described above are combined in this report to provide an integrated local threat index. Past thermal stress (described below) is treated as an additional threat.		
Past thermal stress	Estimates of thermal stress over the past 10 years (1998 to 2007) combine the following two data layers: 1. Past intense heating events. These were areas known to have had high temperature anomalies (scores of degree heating weeks > 8), based on satellite sea surface temperature data provided by NOAA Coral Reef Watch; and 2. Observations of severe bleaching from ReefBase. These point data were buffered to capture nearby bleaching, but modified and effectively reduced by the adjacent presence of low or zero bleaching records from the same year.	■ Estimates of bleaching from remote sensing are a measure of the conditions that may cause bleaching based on the weekly temperatures and long-term averages at the location. ■ Bleaching susceptibility due to other factors (either local or climate-related, such as past climactic variability) was not captured in the model. ■ There is not always a strong correlation between the sea surface temperature and the observations of known bleaching. However, the latter have only a limited spatial and temporal coverage and so cannot be used alone.

Unlike the modeling of local threats, the data and models used to evaluate climate and ocean-chemistry-related threats were obtained from external experts. For this work there were two aims: one to look at recent ocean warming events that may have already degraded reefs or left them more vulnerable to other threats, and the other to project the future impacts from ocean warming and acidification over the medium (20 year) and longer (40 year) term. The stressors for these models include data from satellite observations of sea surface temperature, coral bleaching observations, and modeled estimates of future ocean warming and ocean acidification. Input from scientists from each of the major coral reef regions and from climate change experts contributed to the selection of thresholds for these threats. Table 2.2 summarizes the approach and limitations for the examination of future global-level threats.

The outputs from these models were further tested and calibrated against available information on coral reef condition and observed impacts on coral reefs. All threats were categorized into low, medium, and high threat, both to simplify and to enable fair and direct comparison and to combine findings for the different threats.

TABLE 2.2 *REEFS AT RISK REVISITED* ANALYSIS METHOD—FUTURE GLOBAL-LEVEL THREATS

Threat	Analysis Approach	Limitations
Future thermal stress	■ Projected thermal stress in the 2030s and 2050s is based on modeled accumulated degree heating months (DHM) and represents a "business-as-usual" future for greenhouse gas emissions. ■ The specific indicator used in the model was the frequency (number of years in the decade) that the bleaching threshold is reached at least once. The frequencies were adjusted to account for historical sea surface temperature variability.	■ Data represent a rough approximation of future threat due to thermal stress. ■ Models provide an approximation of a potential future, but variations in emissions and other factors will undoubtedly influence the outcome. ■ Besides historical temperature variability, the model does not incorporate other factors that may induce or prevent coral bleaching (for example, local upwelling, species type), or potential adaptation by corals to increased sea temperatures.
Ocean acidification	■ The indicator of ocean acidification is the projected saturation level of aragonite, the form of calcium carbonate that corals use to build their skeletons. (As dissolved CO_2 levels increase, the aragonite saturation state decreases, which makes it more difficult for coral to build their skeletons.) Aragonite saturation levels were modeled for the future according to projected atmospheric CO_2 and sea surface temperatures levels for 2030 and 2050 based on a "business-as-usual" scenario.	■ Data represent a rough approximation of present and future aragonite saturation levels. ■ Aragonite saturation is an important factor influencing growth rates, but it is likely not the only factor. Other factors (such as light and water quality) were not included in this model due to a lack of global spatial data.

Appendix 2 provides a list of the data sources used in the analysis and the details of model validation. A list of data contributors and full technical notes for the analysis, including data sources and thresholds used to distinguish low, medium and high categories for each threat, are available online at www.wri.org/reefs. Results of the threat analysis are presented in chapters 4 and 5.

Integrating Threats

To develop a single broad measure of threat, we combined the four individual threats to coral reefs into a single integrated local threat index that reflects their cumulative impact on reef ecosystems. We then adjusted this index by increasing threat levels to account for the impacts of past thermal stress. Finally, we combined the local threats with modeled future estimates of thermal stress and ocean acidification to predict threat to reefs in 2030 and 2050.

a. **Integrated Local Threat Index.** This index was developed by summing the four individual local threats, where reefs were categorized into low (0), medium (1), or high (2) in each case. The summed threats were then categorized into the index as follows:[a]

- *Low:*[b] 0 points (scored low for all local threats)
- *Medium:* 1–2 points (scored medium on one or two local threats or high on a single threat)
- *High:* 3–4 points (scored medium on at least three threats, or medium on one threat and high on another threat, or high on two threats)
- *Very high:* 5 points or higher (scored medium or higher on at least three threats, and scored high on at least one).

The resulting integrated local threat index is the most detailed output from the model and is presented on the map inside the front cover and on regional maps in Chapter 5.

(Maps of individual threats are also available online at www.wri.org/reefs.)

b. **Integrated Local Threat and Past Thermal Stress Index.** Thermal stress can cause coral bleaching even on otherwise healthy reefs. When it coincides with local threats, it serves as a compounding threat. To reflect the pressure of thermal stress and local threats, we combined the integrated local threat index with data indicating locations of severe thermal stress events between 1998 and 2007. Reefs in areas of thermal stress increased in threat by one level.[24] These results are presented in the threat summary (Figures 4.6, Table 4.1, and Figures 5.1–5.6) in chapters 4 and 5.

a. Several integration methods were evaluated. This method was chosen because it had the highest correlation with available data on coral condition. The index is slightly more conservative than the previous *Reefs at Risk* reports where a "high" in a single threat would set the integrated local threat index to high overall.

b. The default threshold is "low" when a coral reef is not threatened by a specific local threat. Thus, all reefs are assigned a threat level. This approach assumes that no reef is beyond the reach of human pressure.

c. **Integrated Local Threat and Future Global-Level Threat Index.** We combined the integrated local threat index with modeled projections of ocean acidification and thermal stress in 2030 and 2050 (Table 2.2) to estimate the future threats to coral reefs from climate change. In combining these threats, we weighted local threats more heavily, in light of the greater uncertainty associated with future threats, and the finer resolution of local threat estimates. Reefs are assigned to their threat category from the integrated local threat index as a starting point. Threat is raised one level if reefs are at high threat from either thermal stress or ocean acidification, or if they are at medium threat for both. If reefs are at high threat for both thermal stress and acidification, the threat classification is increased by two levels. In order to portray some nuance in the degree of threat, we have extended the rating scale to include one additional threat category above very high called "critical." The analysis assumes no change in current local threat levels, either due to increased human pressure on reefs or changes in reef-related policies and management. The results of this analysis are presented in Figure 4.9 and Maps 4.2a, b, and c in chapter 4.

LIMITATIONS

The analysis method is of necessity a simplification of human activities and complex natural processes. The model relies on available data and predicted relationships, but cannot capture all aspects of the dynamic interactions between

BOX 2.1. ASSESSING CORAL REEF CONDITION IN THE WATER

A unique and important feature of the *Reefs at Risk* approach is its global coverage—assessing threats to all reefs, even those far from human habitation and scientific outreach. It is, however, a model, and it measures threat rather than condition. Some threatened reefs may still be healthy, but many others will have already suffered some level of degradation. The only way to accurately assess condition is through direct measurement of fish, benthic cover (live coral, dead coral, algae, etc.), or other characteristics. Some reefs, including the Great Barrier Reef, have detailed and regular surveys covering numerous areas, but for most of the world such observation or monitoring is sparse and irregular. There are, however, many national and several international monitoring programs that provide important information, improving our understanding of coral conditions and trends.

- Reef Check is a volunteer survey program with sites in over 80 countries and territories worldwide.
- The Global Coral Reef Monitoring Network (GCRMN) is a network of scientists and reef managers in 96 countries who consolidate status information in periodic global and regional status reports.
- The Atlantic and Gulf Rapid Reef Assessment (AGRRA) Program is a standardized assessment method that has been applied in over 800 coral reef locations across the Caribbean and Gulf of Mexico.
- The Australian Institute of Marine Science (AIMS) conducts scientific research on all of Australia's reefs, including a long-term monitoring program that has been surveying the health of 47 reefs on the Great Barrier Reef annually since 1993.
- Le Centre de Recherches Insulaires et Observatoire de l'Environnement (CRIOBE) conducts periodic monitoring of coral and fish stocks in the South Pacific.
- Coastal Oceans Research and Development in the Indian Ocean (CORDIO) monitors trends in coral reef health, fish populations, and coastal resources in 19 countries in the central and Western Indian Ocean.
- The Reef Environmental Education Foundation (REEF) works with volunteer divers to collect data on marine fish populations in the Caribbean and Pacific.

Visit www.wri.org/reefs for more information about coral reef monitoring programs.

people, climate, and coral reefs. Climate change science, in particular, is a relatively new field in which the complex interactions between reefs and their changing environment are not yet fully understood.

The threat indicators gauge current and potential risks associated with human activities, climate change, and ocean acidification. A strength of the analysis lies in its use of globally consistent data sets to develop globally consistent indicators of human pressure on coral reefs. We purposefully use a conservative approach to the modeling, where thresholds for threat grades are set at reasonably high levels to both counter any data limitations and avoid exaggerating the estimated threats.

The *Reefs at Risk Revisited* analysis is unique in its global scope and ability to provide a big-picture view of threats to reef condition. However, the model is not perfect, and omissions and other errors in the data are unavoidable. For example, the modeling did not include the potentially compounding threats of coral disease or increased storm intensity because of too many uncertainties in their causes, distribution, and relationships. However, a map of global observations of coral diseases can be found in chapter 3 (Map 3.5).

Monitoring data and expert observations were used, where available, to calibrate the individual threat layers and validate the overall model results. The thresholds chosen to distinguish low, medium, and high threat rely heavily on the knowledge of project collaborators with expertise across regions and aspects of reefs and reef management. Their review of model results also served as our most comprehensive validation of results. (Appendix 2 lists collaborators who contributed data or advised on modeling methods.)

PHOTO: WOLCOTT HENRY

Chapter 3. THREATS TO THE WORLD'S REEFS

Coral reefs are fragile ecosystems that exist in a narrow band of environmental conditions. Corals thrive in clear, warm waters that are low in nutrients and have abundant light to support the photosynthetic activities of the symbiotic algae (zooxanthellae) that flourish within coral tissues and are critical to growth. Reefs are also extremely vulnerable to overexploitation. Removal of key functional elements of reef ecosystems, such as larger predators or grazing fish, can have far-reaching consequences across the entire ecosystem.

The local and global threats to coral reefs are described in greater detail in the following sections, which are structured around the source or driver of the threat. For each category of threat, we provide a description of the threat and suggest some options for mitigation.

LOCAL THREATS

COASTAL DEVELOPMENT

Description of threat: Some 2.5 billion people—nearly 40 percent of the global population—live within 100 km of the coast.[25] Development in the coastal zone—linked to human settlements, industry, aquaculture, or infrastructure—can have profound effects on nearshore ecosystems. Impacts of coastal development on the reef can occur either through direct physical damage such as dredging or land filling, or indirectly through increased runoff of sediment, pollution, and sewage.

Large quantities of sediments can be washed into coastal waters during land clearing and construction. The removal of coastal vegetation, such as mangroves, also takes away a critical sediment trap that might otherwise prevent damage to nearshore ecosystems.

Where coastal areas are developed, pollution often follows. Sewage is the most widespread pollutant, and elevated nutrient levels present in sewage encourage blooms of plankton that block light and encourage growth of seaweeds that compete for space on the reef.[26] Many countries with coral reefs have little to no sewage treatment; the Caribbean, Southeast Asia, and Pacific regions discharge an estimated 80 to 90 percent of their wastewater untreated.[27] Toxic chemicals also are a problem. Sources of toxic chemicals in coastal runoff include industries, aquaculture, and agriculture, as well as households, parking lots, gardens, and golf courses.

Direct construction within the marine environment can have even more profound effects. In many areas, wide shallow expanses of reef flats have been targeted for reclamation, and converted to airports or industrial or urban land.

BOX 3.1 REEF STORY

Guam: Military Development Threatens Reefs

The United States recently proposed plans to expand military operations on the U.S. territory of Guam with the construction of new bases, an airfield, a deep-water port, and facilities to support 80,000 new residents (a 45 percent increase over the current population). Dredging the port alone will require removing 300,000 square meters of coral reef. In February 2010, the U.S. Environmental Protection Agency rated the plans as "Environmentally Unsatisfactory" and suggested revisions to upgrade existing wastewater treatment systems and lessen the proposed port's impact on the reef. At the time of publication, construction had not started pending resolution of these issues. *See full story online at www.wri.org/reefs/stories.*

Story provided by Michael Gawel of the Guam Environmental Protection Agency.

PHOTO: LAURIE RAYMUNDO

Elsewhere, the dredging and construction associated with building ship ports and marinas have directly impinged upon reefs. Even coastal engineering in waters adjacent to reefs can alter water flows and introduce sediments, with effects far beyond the location of the construction site.

In some cases, tourism can also threaten reefs. Hotels can bring coastal development to new and remote locations, with associated higher levels of construction, sewage, and waste. Meanwhile, tourists stimulate demand for seafood and curios, beachgoers may trample nearshore reefs, and inattentive recreational divers can break fragile corals. Damaged corals then become more susceptible to disease and algal overgrowth.[28]

Trends: Population growth in coastal areas continues to outpace overall population growth. Between 2000 and 2005, population within 10 km of the coast grew roughly 30 percent faster than the global average.[29] As populations grow and natural ecological buffers on the shoreline are lost, sea level rise and changing storm patterns due to climate change are likely to lead to increased coastal engineering activities for seawalls and other mitigating construction.

Remedies: The impacts of coastal development can be greatly reduced through effective planning and regulations. These include zoning regulations, protection of mangroves and other vegetation, and setbacks that restrict development within a fixed distance of the coastline.[30] For example, any new development in Barbados must be 30 meters behind the high water mark.[31] Such precautions also prevent the need for future coastal engineering solutions by allowing for the natural movements of beaches and vegetation over time, thus saving future costs and unintended consequences. Where land-filling or harbor development is deemed necessary, methods for reducing impacts in adjacent waters include using silt fences, settling ponds, or vegetated buffer strips to trap sediments before they enter waterways.[32]

Improvement in the collection and treatment of wastewater from coastal settlements benefits both reefs and people through improved water quality and reduced risk of bacterial infections, algal blooms, and toxic fish. Estimates show that for every US$1 invested in sanitation, the net benefit is US$3 to US$34 in economic, environmental, and social improvement for the nearby community.[27]

Pressure from tourism can be reduced through proper siting of new structures, including measures such as honoring coastal setbacks; retaining mangroves and other coastal habitats; using environmentally sound materials (for example, avoiding sand and coral mining); and installing and maintaining effective sewage treatment. Managing tourism within sustainable levels is also important, such that visitation does not degrade the reef. Educated tourists help to create a demand for responsible coastal development.

WATERSHED-BASED POLLUTION

Description of threat: Human activities far inland can impact coastal waters and coral reefs. As forests are cut or

BOX 3.2 PHOTO STORY—DISAPPEARING MANGROVES

Mangroves line the coast in many coral reef regions. They provide a critical buffer, holding back sediments washed from the land as well as reducing nutrients and other pollutants.[33] Pressure for coastal development, including conversion to agriculture and aquaculture, has led to rapid losses of mangroves—nearly 20 percent have disappeared since 1980.[34] With the loss of mangroves, reefs are more vulnerable to pollution from the land. There may also be more direct ecological impacts through the many reef creatures that utilize mangroves as a nursery area, or as a valuable adjacent habitat for feeding, hiding, or breeding.[35]

pastures plowed, erosion adds millions of tons of sediment to rivers, particularly in steeper areas and places with heavy rainfall. Agriculture adds more than 130 million tons of fertilizer (i.e., nutrients) and pesticides worldwide to crops each year,[36] much of which enters waterways where they are transported to the coast.[37] Livestock can compound these problems. Overgrazing removes vegetation and adds to erosion, even on many uninhabited islands with populations of feral sheep or goats. Meanwhile, livestock waste adds considerable nutrient pollution to waterways leading to the sea. Mining also represents a more localized threat through sediment runoff or leaching of chemical toxins.

At the coast, sediments, nutrients, and pollutants disperse into adjacent waters, some plumes reaching more than 100 km from the river mouth.[38] Such impacts can be reduced where mangrove forests or seagrass beds lie between the rivers and the reefs. Both of these habitats can help to trap sediments as they settle out in the calm waters among shoots and roots, and can also play a role in the active removal of dissolved nutrients from the water.[34, 39]

In high quantities, sediments can smother, weaken, and kill corals and other benthic organisms. In lower quantities, they reduce the ability of zooxanthellae to photosynthesize, slowing coral growth.[32] Excessive levels of nutrients like nitrogen and phosphorus in shallow coastal waters (that is, eutrophication) can encourage blooms of phytoplankton in the water, which block light from reaching the corals, or they can cause vigorous growth of algae and seaweeds on the sea bed that out-compete or overgrow corals.[40] In severe cases, eutrophication can lead to hypoxia, where decomposition of algae and other organisms consumes all of the oxygen in the water, leading to "dead zones" and eventually nearshore ecosystem collapse.[41]

Trends: Deforestation is a major contributor of sediment to watersheds. Between 2000 and 2005, an estimated 2.4 percent of humid tropical forests were lost to deforestation, with some of the most intense clearing occurring in the coral reef countries of Brazil, Indonesia, Malaysia, Tanzania, Myanmar, and Cambodia.[42] Meanwhile, climate change is expected to cause heavier and more frequent precipitation in many areas, which would exacerbate pollution and sediment runoff to the coast.[43]

To support the food demands of a growing global population, agriculture will increase both in extent and intensity. The FAO estimates that total fertilizer use will grow approximately 1 percent per year—from a baseline of 133 million tons per year in 1997 to 199 million tons per year in 2030.[36] Developing countries, notably in Africa and South Asia, are expected to have the highest growth rates in fertilizer consumption. Hypoxia is a growing problem in coastal waters, where the number of documented cases worldwide grew from 44 in 1995 to 169 in 2007.[41]

Remedies: Land management policies and economic incentives are important for reducing watershed-based threats. Improved agricultural methods can both reduce erosion and runoff, as well as increase fertilizer efficiency, benefiting both farmers and fishers. For example, conservation tillage (leaving previous vegetation untilled in the soil) helps to reduce both soil loss and farmer labor and fuel expenditures, while contour plowing or the use of terraces reduces

BOX 3.3 REEF STORY

Palau: Communities Manage Watersheds and Protect Reefs

The Republic of Palau, in the western Pacific Ocean, is surrounded by more than 525 sq km of coral reefs. Construction of the recently completed 85-km "Compact Road" around Palau's largest island, Babeldaob, led to widespread clearing of forests and mangroves, causing soils to erode into rivers and coastal waterways, damaging coral reefs, seagrass beds, and freshwater resources. To better understand the impact of the changing landscape on the marine environment, the Palau International Coral Reef Center (PICRC) conducted a study that revealed that the degradation of reefs was a direct result of land-based sediments. After PICRC presented these findings to local communities, the governing body of Palau's Airai State instituted a ban on the clearing of mangroves. Communities, local governments, and NGOs also joined together to form the Babeldaob Watershed Alliance, a forum for developing land management plans and establishing collective conservation goals. *See full story online at www.wri.org/reefs/stories.*

Story provided by Steven Victor of the The Nature Conservancy, Palau.

erosion. Nutrient runoff can be reduced by pre-testing soils for nutrient levels and improving the timing of fertilizer applications. Agroforestry, reforestation, protection of forests on steep slopes, and requiring forest belts to be left along river margins can all greatly reduce the release of nutrients and sediments into waterways and improve the reliability of year-round freshwater supplies. Preservation of wetlands, mangroves, and seagrasses at the coast can filter and trap sediments and nutrients before reaching reefs.

MARINE-BASED POLLUTION AND DAMAGE

Description of threat: Thousands of commercial, recreational, and passenger vessels pass near reef areas every day, bringing with them a host of potential threats, including contaminated bilge water, fuel leakages, raw sewage, solid waste, and invasive species. In addition, reefs are exposed to more direct physical damage from groundings, anchors, and oil spills.

Marine-based sources of pollution can rapidly undermine the health of coral reefs. For example, oil from spills, leaks in rigs and pipelines, or from ship discharge can have both short-term and long-term (chronic) effects. Studies on corals exposed to oil have identified tissue death, change in calcification rate, expulsion of zooxanthellae, and larval death among other serious stress responses.[44]

Cruise ships are a significant source of pollution in many areas. In 2009, more than 230 cruise ships hosted an estimated 13.4 million passengers worldwide.[45] A typical one-week cruise on a large ship (3,000 passengers and crew) generates almost 800 cubic meters of sewage; 3,700 cubic meters of graywater; half a cubic meter of hazardous waste; 8 tons of solid waste; and nearly 100 cubic meters of oily bilge water.[45] Estimates suggest that a typical cruise ship generates 70 times more solid waste per day than a typical cargo ship.[31] The International Convention for the Prevention of Pollution from Ships (MARPOL) provides a set of approved guidelines regulating the discharge of sewage, oily bilge water, hazardous wastes, and solid waste (which includes a ban on all dumping of plastics). Unfortunately, MARPOL's regulations are met with varying degrees of compliance within the cruise industry and beyond.

Invasive species—accidentally transported from distant locations in the ballast water of ships or released from aquariums—also impact coral communities by killing off or displacing native species.[46] Examples in tropical waters include lionfish, a native of the Indo-Pacific now found throughout the Caribbean, and several types of invasive algae in the Hawaiian Islands.[47] Reefs located near ports of call are most at risk from invasive species. It has been estimated that, at any one time, as many as 10,000 marine species may be transported globally in ships' ballast water,[48] though only a tiny fraction of these survive the trip or colonize a new location.

Ships and other vessels can also be a source of direct physical damage and destruction to reefs. Contact with ship hulls, anchors, or propellers can crush, break, or dislodge corals. Smaller vessels generally cause lighter damage, but the cumulative impact can be dramatic in areas of heavy recreational boating, such as the Florida Keys National Marine Sanctuary, which records 60 to 90 groundings on reefs annually, though many more likely go unreported.[49] Marine debris from ships, including plastics and abandoned fishing gear, can also cause physical damage to reefs and entangle marine organisms such as fish and turtles.[50] It can take corals decades to recover from physical damage caused by boat strikes and marine debris.[51]

Trends: Despite growing efforts to regulate greenhouse gas emissions, global demand for oil is increasing and is expected to grow from 83 million barrels per day in 2004 to 118 million barrels per day by 2030.[52] While techniques to avoid spillage and loss have improved, so have the net risks of spillage, given the continuing increases in volume and the increasingly challenging environments from which oil is drilled. A prime example of this risk is the 2010 Deepwater Horizon oil spill in the Gulf of Mexico, where inadequate government oversight and a failure to follow precautionary measures contributed to one of the largest marine oil spills in the history of the United States.[53]

Maritime shipping continues to grow rapidly compared to other forms of transportation. Estimated gross tonnage of international commercial shipping increased by 67 percent between 1980 and 2003.[54] Cruise tourism also continues to grow. The number of cruise passengers has increased by an average of 7.4 percent per year since 1980 and 118 new ships have been launched since 2000.[45] In terms of waste management, the cruise industry is generally improving. Some ships now have advanced sewage treatment, shipboard recycling programs, and increased use of biodegradable alternatives to plastics.[55] As maritime transport continues to grow, however, the threat posed by invasive species also increases since the threat of accidental release from ballast water or biofouling on ships' hulls is difficult to manage.

BOX 3.4 REEF STORY

American Samoa: Shipwreck at Rose Atoll National Wildlife Refuge

Rose Atoll is a National Wildlife Refuge located in the South Pacific within the U.S. territory of American Samoa. In 1993, a 275-ton fishing vessel ran aground on Rose Atoll's shallow reef. Initially, only the bow section of the ship was removed. However, subsequent monitoring revealed that the disintegration and corrosion of the ship was releasing dissolved iron into surrounding waters, stimulating growth of blue-green algae on the reefs. In response, the U.S. government removed the remaining debris at a substantial cost. The reefs are now recovering rapidly. This success was due largely to Rose Atoll's status as an actively managed protected area, in combination with sufficient funds, effort, and expertise to monitor the damage and recovery. *See full story online at www.wri.org/reefs/stories.*

Story provided by James Maragos of the US Fish and Wildlife Service, Hawaii.

PHOTO: U.S. FISH AND WILDLIFE SERVICE

Remedies: Environmental control measures at the local level are essential for mitigating marine-based pollution and damage to reefs. Such measures include developing infrastructure at ports to accept and properly dispose of ship-generated waste; improving wastewater treatment systems on cruise ships and cargo ships; restricting shipping lanes to route traffic away from reefs; and developing effective oil spill contingency plans. Ballast water regulations, which require the disposal or exchange of ballast water far offshore in deep waters before ships can enter ports, are important for reducing the risk of invasive species entering coastal waters. Expanding the availability of fixed moorings for recreational craft can reduce anchor damage and the likelihood of groundings. Educating vessel owners can also help with compliance.

BOX 3.5 REEF STORY

Tanzania: Deadly Dynamite Fishing Resurfaces

Tanzania, on Africa's east coast, is home to an extensive network of coral reefs that support major artisanal fishing and tourism industries. However, Tanzania is also the only country in Africa where dynamite fishing still occurs on a large scale. This devastating form of fishing first appeared in the 1960s, and by the mid-1990s had become a serious problem. A high-profile national campaign in the late 1990s nearly eradicated blast fishing between 1997 and 2003; however, inadequate prosecution and minimal penalties levied against dynamiters have allowed this illegal practice to re-emerge and expand. Increased pressure, both domestically and internationally, is needed to create the political will necessary to once again halt this short-sighted and unsustainable practice. *See full story online at www.wri.org/reefs/stories.*

| *Story provided by Sue Wells (Independent).*

Adopting and enforcing national legislation in all coral reef countries to incorporate international agreements on marine pollution would greatly help to reduce marine-based threats to reefs. Besides MARPOL, other International Maritime Organization (IMO) treaties include the London Convention and Protocol and the International Convention on Oil Pollution Preparedness, Response, and Cooperation (OPRC), which address waste disposal and oil spills at sea, respectively. Even tighter regulation on oil exploration and exploitation in challenging environments such as deepwater areas may also be needed to prevent future catastrophic oil spills.

OVERFISHING AND DESTRUCTIVE FISHING

Description of threat. Reef fisheries have long sustained coastal communities by providing sources of both food and livelihoods. However, over 275 million people currently live within 10 km of the coast and 30 km of a coral reef,[56] and fishing pressure is high on many reefs. When well-managed, such fisheries can be a sustainable resource, but growing human populations, more efficient fishing methods, and increasing demands from tourism and international markets have significantly impacted fish stocks. Some target species—including groupers, snappers, sharks, sea cucumbers and lobsters—command such high prices as export commodities that fishing vessels travel hundreds, even thousands, of kilometers to reach the last remote strongholds and often fish illegally in protected or foreign waters to secure catches. [57, 58]

Removing just one group of fish from the reef food web can have cascading effects across the ecosystem.[18] If top predators are taken, prey species are no longer held in check, and the overall response can be both complex and somewhat unpredictable, potentially causing an overall destabilization of the system. While large predators are often preferred target species, as their numbers decline, fishers move to smaller, often herbivorous fish (in a process known as "fishing down the food chain").[59] Heavily fished reefs are thus left with low numbers of mostly small fish. Such reefs are then prone to algal overgrowth, without herbivores to graze the algae as they grow. Such overfished reefs appear to be generally less resilient to stressors, and may be more vulnerable to disease and slower to recover from other human impacts.[60, 61, 62]

In some places, the fishing methods themselves are destructive. Most notable is the use of explosives to kill or stun fish, which destroys coral in the process.[63] Although illegal in many countries, blast (or dynamite) fishing remains a persistent threat, particularly in Southeast Asia and East Africa.[64] Poison fishing is also destructive to corals. This practice typically involves using cyanide to stun and capture fish live for the lucrative live reef food fish or aquarium fish markets. The poison can bleach corals and kill polyps. Fishers often break corals to extract the stunned fish, while other species in the vicinity are killed or left vulnerable to predation.[65] Map 3.1 provides a summary of locations identified as threatened by fishing with explosives or poison.

MAP 3.1. GLOBAL OBSERVATIONS OF BLAST AND POISON FISHING

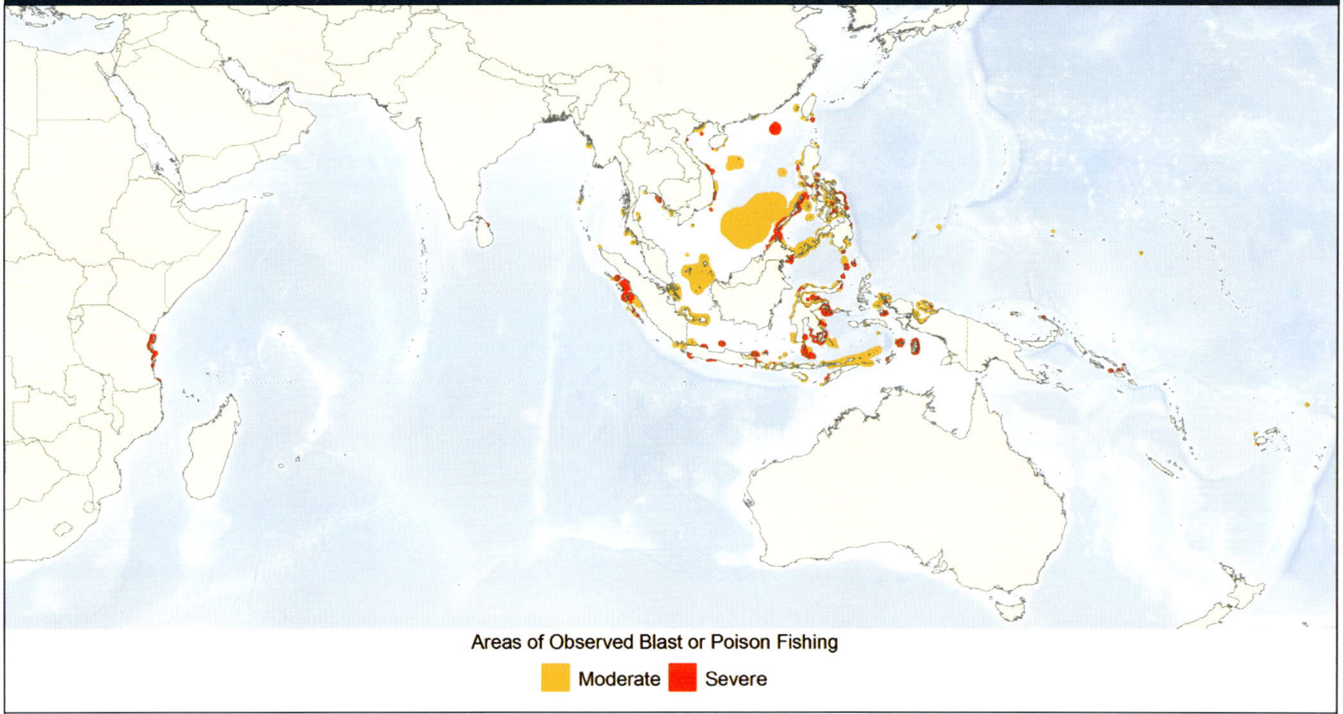

Note: Blast and poison fishing is largely undertaken in Southeast Asia, the western Pacific, and eastern Africa. Areas of threat shown here are based on survey observations and expert opinion.
Source: WRI, 2011.

Certain types of fishing gear can also have a destructive impact on a reef ecosystem. Gill nets and beach seines drag along the sea bottom, capturing or flattening everything in their path, including non-targeted or juvenile species and delicate corals. Furthermore, discarded or lost nets and traps can continue "ghost fishing"—ensnaring prey and smothering corals—for months or years after their original use.

Trends: Important drivers of unsustainable fishing include population growth, excess fishing capacity, poor fisheries governance and management practices, international demand for fish, and a lack of alternative income options in coastal communities. Globally, the Food and Agriculture Organization of the United Nations (FAO) estimates that 80 percent of the world's wild marine fish stocks are fully exploited or overexploited.[66] These numbers do not consider the impact of illegal, unreported, and unregulated catches, which were estimated to add approximately 20 percent to official catch statistics between 1980 and 2003.[67] Most coral reef fisheries are small-scale fisheries, and thus are poorly represented in global fisheries statistics.[68, 69] However, where national figures are available, such as for Indonesia and the Philippines, they indicate severe problems.[7, 16] Unsustainable fishing of some species is reported even on some of the most remote and best-protected coral reefs.[57] Thus it is highly likely that most reef fisheries around the world are in similar or worse condition than indicated by FAO's global assessments.

On the positive side, national and local governments have designated an increasing number of marine protected areas (MPAs) in an effort to protect reefs. These include many sites in areas where human pressures are considerable. Such sites, especially where they have community support, can be remarkably effective in reducing fishing pressure. However, sites in high-pressure areas still make up only a very small proportion of reefs, and the largest MPAs tend to be more remote. A number of very large MPAs have greatly added to global coverage, including Papahānaumokuākea Marine National Monument in the Northwestern Hawaiian Islands, which spans 360,000 sq km (an area roughly the size of Germany); [70] the Phoenix Islands Protected Areas, which cover 408,250 sq km of the mostly uninhabited Phoenix Islands and surrounding waters; and the recently

BOX 3.6. TAKEN ALIVE—FISH FOR AQUARIUMS AND THE LIVE REEF FOOD FISH TRADE

Both the live reef food fish—that is, fish captured to sell live in markets and restaurants—and the ornamental species trades are high-value industries. The ornamental species trade takes in an estimated $200 million to $330 million per year globally, with the majority of exports leaving Southeast Asian countries and entering the United States and Europe. The overall value of the industry has remained stable within the past decade, though trade statistics are incomplete.[76] The live reef food fish trade is concentrated mainly in Southeast Asia, with the majority of fish exported from the Philippines and Indonesia and imported through Hong Kong to China.[77] Over time, the trade has expanded its reach, drawing exports from the Indian Ocean and Pacific islands, reflecting depleted stocks in Southeast Asia, rising demand, improvements in transport, and the high value of traded fish.[16] The estimated value of the live reef food fish trade was $810 million in 2002.[78] A live reef food fish sells for approximately four to eight times more than a comparable dead fish, and can fetch up to $180 per kilogram for sought-after species like Napoleon wrasse or barramundi cod, making it a very lucrative industry for fishers and traders alike.[77]

PHOTO: JULIE MCGOWAN, TIMANA PHOTOGRAPHY, 2006/MARINE PHOTOBANK

designated Chagos Archipelago MPA, which covers approximately 550,000 sq km.[71] The strength of regulations for fisheries across MPA sites globally is variable, but "no-take" reserves (where all fishing is banned) form an important part of the mix. These include zones within MPAs as well as entire MPAs. For example, the area designated as no-take in the Great Barrier Reef Marine Park increased from less than 5 percent to 33 percent in 2004, equaling over 115,000 sq km, and has already had dramatic positive benefits on the reef.[72, 73, 74, 75]

Remedies: Fisheries management can take many forms, including seasonal closures to protect breeding sites; restrictions on where and how many people are allowed to fish; and restrictions on the sizes or quantities of fish they can take or on the types of fishing gear they can use.[79] Areas closed to fishing can show rapid recovery, with more and larger fish within their boundaries, associated benefits for corals and other species, and "spillover" of adult fish stocks at the perimeter that can enhance fisheries in adjacent areas.[74, 80, 81] In all cases, size and placement are important for achieving success. Enforcement is critical, and local support and community involvement in management are essential for effective management.

Many countries already have laws against blast and poison fishing, but need to apply more resources toward enforcement. Countries could also regulate the import and export of live food fish and aquarium fish to ensure they were caught using sustainable and nondestructive fishing methods.[82] At the local level, education is an important tool for increasing awareness among fishers that destructive fishing practices negatively impact the very resources that provide their food and livelihoods. There are also growing efforts to encourage a more active role for consumers, and private market agreements in the fish trade worldwide. Certification and ecolabeling, such as that of the Marine Stewardship Council, may help alter market demand and increase premiums paid for fish that are sustainably sourced, although efforts to date have had limited effect on reducing overfishing.[83]

CHANGING CLIMATE AND OCEAN CHEMISTRY

WARMING SEAS

Description of threat: Corals are highly sensitive to changes in temperature. During unusually warm conditions corals exhibit a stress response known as bleaching, in which they lose the microscopic algae (zooxanthellae) that usually live within their tissues. Without zooxanthellae, living coral tissue becomes transparent and the limestone skeleton underneath becomes visible. Depending on the duration and level of temperature stress, coral reefs can either die or survive bleaching. However, even reefs that recover are likely to exhibit reduced growth and reproduction, and may be more vulnerable to diseases.

Natural variation in water temperatures, together with other local stressors, has always caused occasional, small-scale episodes of coral bleaching. Recent years, however, have seen a rise in the occurrence of abnormally high ocean temperatures,[84, 85] which has led to more frequent, more intense, and more widespread "mass bleaching" events where numerous corals of many different species across a large area bleach simultaneously.[86] The most notable mass bleaching event to date occurred in 1998, when wide areas of elevated water temperatures were recorded across many parts of the tropics, linked to an unusually strong El Niño and La Niña sequence (a natural, but dramatic global fluctuation in ocean surface waters and in associated weather patterns). Such events, in combination with the background rate of global warming, can produce particularly high temperatures in some regions .[87] The result in 1998 was that bleaching affected entire reef ecosystems in all parts of the world, killing an estimated 16 percent of corals globally.[47, 88] In the worst-hit areas, such as the central and western Indian Ocean, 50 to 90 percent of all corals died.[89] New coral growth has been variable, but only three-quarters of reefs affected have since recovered (Box 3.7).[47, 90]

Further temperature-driven mass bleaching has occurred since 1998, and in some regions it has caused even greater damage. Extensive bleaching occurred on the Great Barrier Reef in 2002,[91] while 2005 saw the most severe bleaching to date in parts of the Caribbean.[92, 93] Approximately 370 observations of coral bleaching were reported globally between 1980 and 1997, while more than 3,700 were reported between 1998 and 2010 (Figure 3.1). As this report

FIGURE 3.1. TRENDS IN CORAL BLEACHING, 1980–2010

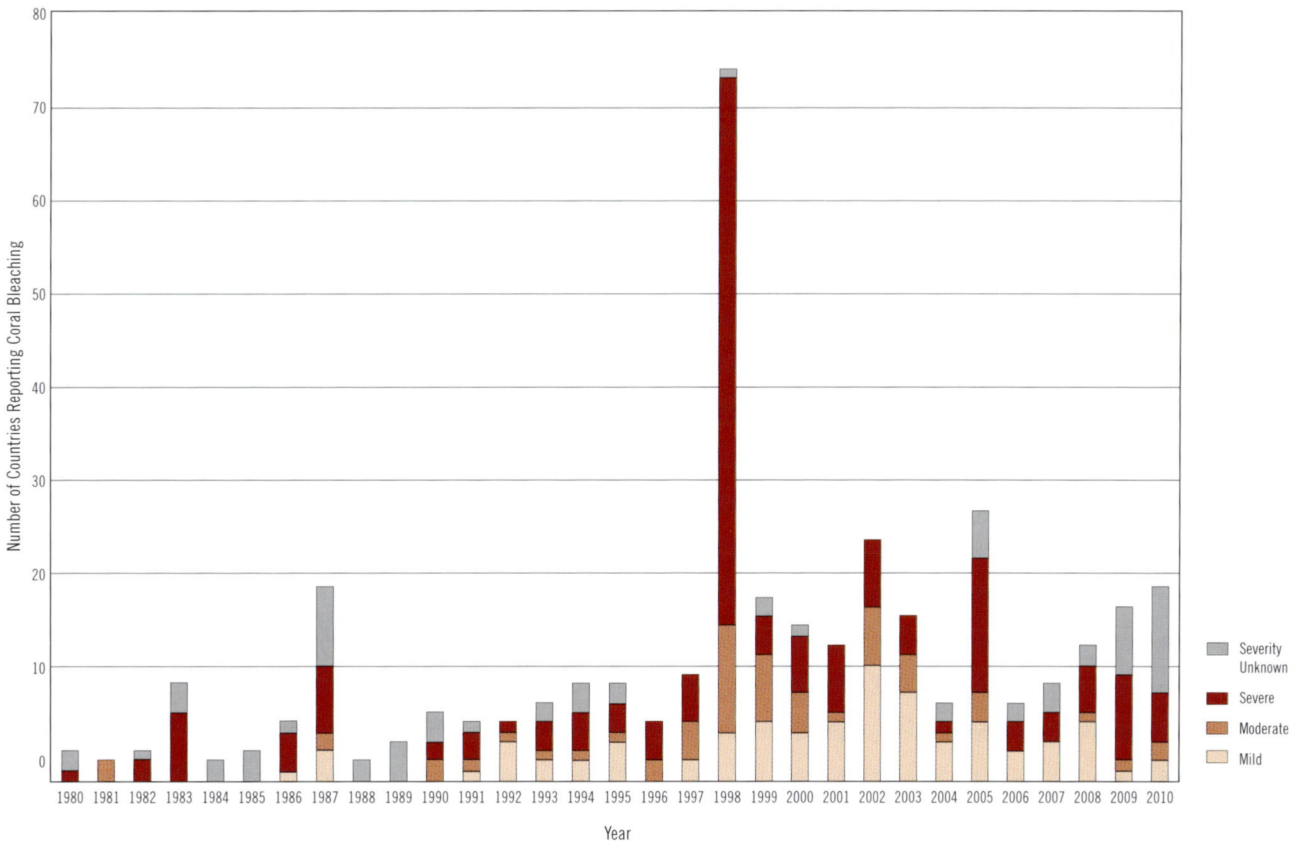

Note: Data for 2010 are incomplete.
Source: ReefBase, 2010.

was being finalized, reports of major bleaching in 2010 were still coming in, but pointed to a mass bleaching event in multiple regions. The increase in recorded observations over time reflects rising sea surface temperatures as well as increased awareness, monitoring, and communication of bleaching events. Map 3.2 shows observed bleaching observations and modeled thermal stress from 1998 to 2007.

Not all corals are equally susceptible to bleaching. Some species appear to be more tolerant, and some individuals appear better acclimated as a result of past exposure to stresses. In all cases, however, such acclimation capacity is limited, and all corals seem to be susceptible to bleaching under the most extreme warming.[94] There appears to be variation in how well different reef communities within an ecosystem survive or recover from bleaching events.[95] This variation may be due to environmental factors such as depth, shading, currents, upwelling, and wave action. Corals and reefs that are better able to avoid or tolerate bleaching are termed "resistant." [96, 97] Corals reefs that can recover to their previous state more quickly after a bleaching event are considered to be"resilient."[96, 98, 99] Factors that appear to improve the resilience of a coral reef include good connectivity to unimpacted or resistant reef areas, enabling coral larvae to move in and reestablish the coral population;[97, 100] abundant herbivore populations to graze on algae, maintaining space on the reef surface for corals to recolonize;[21] and the absence of other local threats such as pollution and sedimentation.[101] Despite the potential for resilience, however, there is already evidence of a growing number of reefs for which recovery has been minimal, even over a decade or longer.[102, 103, 104]

Trends: Mass coral bleaching has occurred multiple times since 1983, increasing in frequency and severity as sea temperatures have risen over time.[84, 113] Predictions based on projected temperatures suggest that severe bleaching will occur with increasing frequency on reefs during the next two to three decades.[86, 114, 115] With current global CO_2 emissions matching or exceeding levels projected under the most pessimistic scenarios of the Intergovernmental Panel on Climate Change (IPCC)[116] and the added challenge of "committed warming" (which would occur even if greenhouse gas emissions today were halted, due to lags in the global climate system), continued mass bleaching events seem almost certain.[117]

For this report, we used the best-available models that combine NOAA's methodology for predicting bleaching episodes with estimates of future sea surface temperature due to climate change to predict the frequency of bleaching episodes in the future. Map 3.3 shows the frequency of Bleaching Alert Level 2 for the decades 2030 to 2039 and 2050 to 2059 based on an IPCC A1B ("business-as-usual") emissions scenario. Note that these estimates have been adjusted to account for historical temperature variability, but have not been adjusted by any other resistance or resilience factors. We have used both recent past bleaching likelihood and future

BOX 3.7 REEF STORY

Mesoamerican Reef: Low Stress Leads to Resilience

The Mesoamerican Reef—the largest continuous reef in the Western Hemisphere—is threatened by overfishing, coastal development, agricultural runoff, and warming seas. In 1998, a mass coral bleaching event caused significant coral mortality on the reef. However, some coral species in areas where the reef and surrounding waters were relatively free of sediment were able to recover and grow normally within two to three years, while corals living with excessive local impacts were not able to fully recover even eight years after the event. This pattern suggests that reducing local threats will also help corals to be more resilient in the face of rising sea temperatures.[105] *See full story online at www.wri.org/reefs/stories.*

Story provided by Annie Reisewitz and Jessica Carilli of the Scripps Institution of Oceanography at the University of California, San Diego.

PHOTO: JASON VALDEZ

MAP 3.2. THERMAL STRESS ON CORAL REEFS (1998 TO 2007)

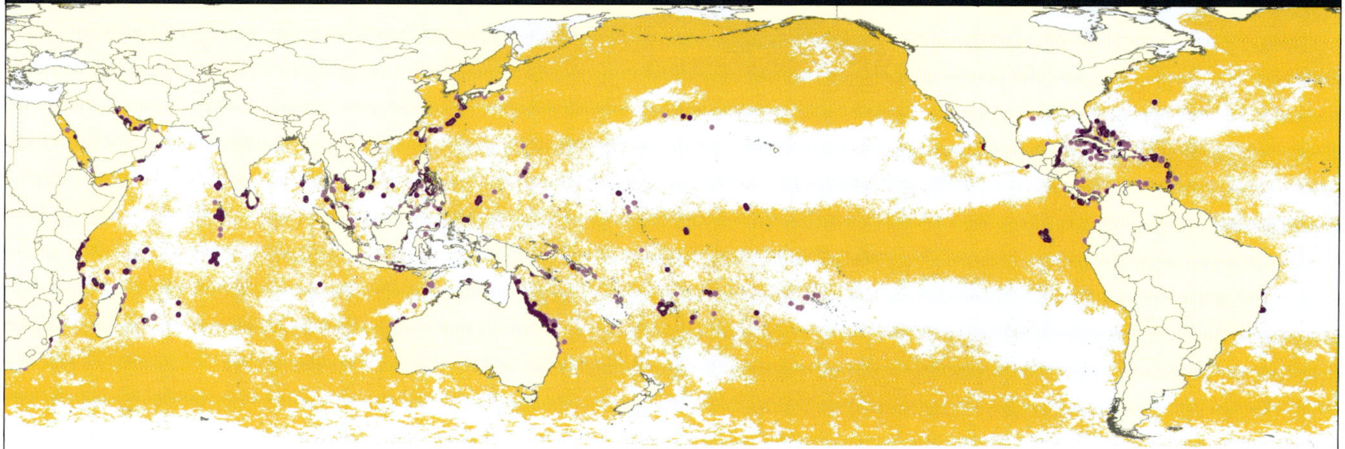

Note: The map reflects the locations of thermal stress on coral reefs between 1998 and 2007 based on coral bleaching observations (in purple) and severe thermal stress from satellite detection (defined as a degree heating week ≥ 8, shown in orange). As many occurrences coral bleaching are unobserved or unreported, we used satellite detected thermal stress as a means of filling in gaps in the observational data.

Source: NOAA Coral Reef Watch, 2010; ReefBase, 2009.

The U.S. National Oceanic and Atmospheric Administration (NOAA) Coral Reef Watch program uses satellites to monitor sea surface temperature (SST) to determine when and where coral bleaching may occur. Their methodology for predicting bleaching is based on abnormally high and sustained SSTs, measured in "degree heating weeks" (DHW), where one DHW is equal to one week of SST 1°C warmer than the historical average for the warmest month of the year. A DHW of 4 (for example, 4 weeks of 1°C warmer or 2 weeks of 2 °C warmer) typically causes widespread coral bleaching and is referred to as a "Bleaching Alert Level 1." A DHW of 8 typically causes severe bleaching and some coral mortality, and is referred to as a "Bleaching Alert Level 2."[109] For this report, high-resolution (~4 km) DHWs were calculated from NOAA's National Oceanographic Data Center Pathfinder Version 5.0 SST data set,[110] using the Coral Reef Watch methodology. Map 3.2 depicts the locations where severe thermal stress ("Bleaching Alert Level 2") was detected by satellite on at least one occasion between 1998 and 2007,[111] along with actual bleaching observations, as recorded in the ReefBase database.[112] These data were combined to assess locations where reefs experienced bleaching-level stress during these years.

Note: A higher DHW threshold was used for the Middle East region (Red Sea and Persian Gulf) to compensate for exaggerated temperature readings driven by land around these enclosed seas. See the full technical notes at www.wri.org/reefs for detailed information on the modification and justification.

bleaching risk maps to develop global-level threat measures for the world's reefs, as described further in chapter 4. Although coral reefs show some capacity to adapt, experts predict that extreme bleaching events could eventually become so frequent that corals will not have time to recover between events.[115] This point may have already been reached in parts of the Caribbean, where bleaching stress is compounded by other local threats and where recovery has been minimal between recent bleaching events.[93]

Remedies: Ultimately the only clear solution to this threat will be a concerted and successful global effort to reduce atmospheric greenhouse gas emissions and to stabilize atmospheric concentrations somewhere around or below current levels.[94]

Recognizing the challenge of global emissions reductions, it is critical that we apply any local measures we can to encourage resistance and resilience. This may buy time for global responses to climate change to take effect, and should help to maintain, for as long as possible, the critical ecosystem services on which so many people depend. A key factor in promoting reef resilience to climate change is the reduction or elimination of local threats. Recommended interventions include reduction in pollution, sedimentation, and overfishing; the protection of critical areas where natu-

MAP 3.3. FREQUENCY OF FUTURE BLEACHING EVENTS IN THE 2030s AND 2050s

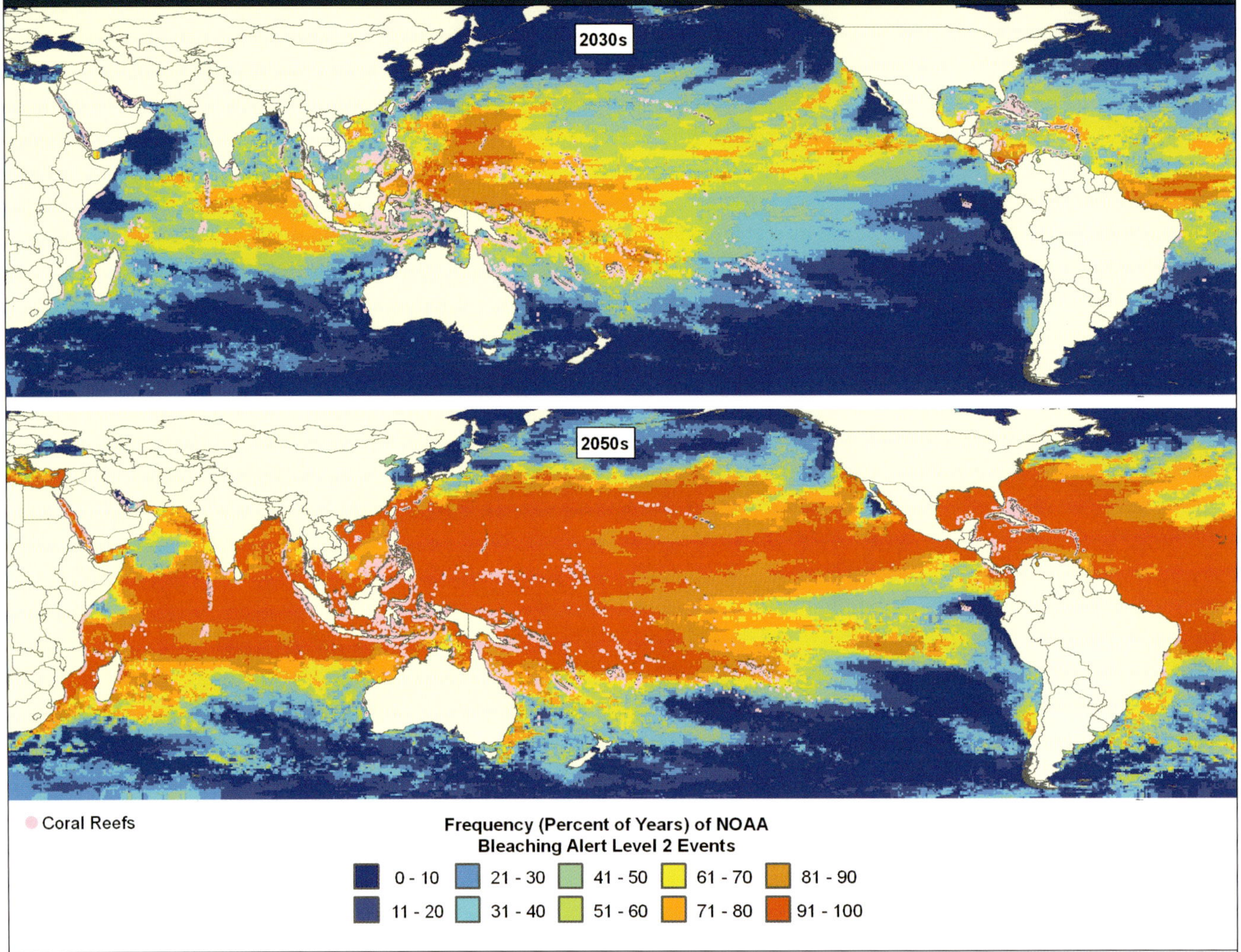

Note: Frequency of future bleaching events in the 2030s and 2050s, as represented by the percentage of years in each decade where a NOAA Bleaching Alert Level 2 is predicted to occur. Predictions are based on an IPCC A1B ("business-as-usual") emissions scenario and adjusted to account for historical temperature variability, but not adjusted by any other resistance or resilience factors.

Source: Adapted from Donner, S.D. 2009. "Coping with commitment: Projected thermal stress on coral reefs under different future scenarios." *PLoS ONE* 4(6): e5712.

ral environmental conditions improve resistance and resilience; the replication of protection such that different reef zones or communities are protected in multiple locations; and rapid adaptive management responses when bleaching events occur.[118] Such responses include measures to reduce local stress on reefs from physical damage (for example, from boats or divers), pollution, or fishing during bleaching events. Past bleaching events have shown that even in the most severe cases, there is rarely a total elimination of corals on a reef, with better survival in certain areas or zones such as areas of local upwelling, localized shading, channels, or lagoon reef patches.[119] Thus, even while research continues to discover the locations of greatest resistance and resilience, and the underlying mechanisms of each, enough is already known to manage reef systems in a way that will encourage resilience.[119] Such measures will not prevent coral bleaching, but they can accelerate recovery.[21]

ACIDIFYING SEAS

Description of threat: In addition to warming the ocean, increases in atmospheric CO_2 will have another impact on coral reefs in coming decades.[120] About 30 percent of the CO_2 emitted by human activities is absorbed into the surface layers of the oceans, where it reacts with water to form carbonic acid.[121] This subtle acidification has profound effects on the chemical composition of seawater, especially on the availability and solubility of mineral compounds such as calcite and aragonite, needed by corals and other organisms to build their skeletons.[20, 122] Initially these changes to ocean chemistry are expected to slow the growth of corals, and may weaken their skeletons. Continued acidification will eventually halt all coral growth and begin to drive a slow dissolution of carbonate structures such as reefs.[123] Such responses will be further influenced by other local stressors. In addition, acidification has also been shown to produce an increased likelihood of temperature-induced coral bleaching.[120] At the ecosystem level, acidification might first affect reefs by reducing their ability to recover from other impacts, and by driving a shift toward communities that include fewer reef-building corals. At the present time, most of the impacts of acidification have been predicted through models and manipulative experiments. However, monitoring on the Great Barrier Reef and elsewhere suggests acidification might already be slowing growth rates.[124] Without significant reductions of emissions, acidification could become a major threat to the continued existence of coral reefs within the next few decades.[124]

Trends: Shallow tropical waters are normally highly saturated with aragonite, the form of calcium carbonate that corals and some other marine organisms use to build their skeletons and shells. However, aragonite saturation levels have fallen dramatically within the past century, from approximately 4.6 to 4.0.[94, 125] An aragonite saturation level of 4.0 or greater is considered optimal for coral growth, while a level of 3.0 or less is considered extremely marginal for supporting coral reefs.[123] These delineations are based on current-day reef distributions, and are thus somewhat subjective, but recent work appears to support this assessment.[20] Map 3.4 compares estimated aragonite saturation states in tropical waters around the world for CO_2 stabilization levels of 380 ppm, 450 ppm, and 500 ppm. These CO_2 stabilization levels correspond approximately to the years 2005, 2030, and 2050 under the IPCC A1B (business-as-usual) emissions scenario.

Scientists have predicted that at CO_2 levels of about 450 ppm, aragonite saturation levels will decrease enough in many parts of the world that coral growth will be severely

BOX 3.9 REEF STORY

Papua New Guinea: Marine Protection Designed for Reef Resilience in Kimbe Bay

Located off the island of New Britain, Papua New Guinea, the rich marine habitat of Kimbe Bay supports local economic and cultural life. However, Kimbe Bay's reefs are particularly threatened by land pollution, overfishing, and bleaching. In response, local communities and government agencies are working together with The Nature Conservancy to design and implement one of the first marine protected area (MPA) networks that incorporates both socioeconomic considerations and the principles of coral reef resilience to climate change, such as biological connectivity (to promote the exchange of larvae between reefs). The lessons learned from this pilot MPA will help to give coral reefs and associated ecosystems around the world a better chance to survive climate change. *See full story online at www.wri.org/reefs/stories.*

Story provided by Susan Ruffo and Allison Green of The Nature Conservancy.

PHOTO: AISON GREEN

BOX 3.8. THE THREAT OF EXTINCTION

Losses of coral reef ecosystem function and of provisioning services typically precede the complete loss of species, but for some species global extinction remains a real risk. IUCN has a formal and consistent framework for assessing extinction risk, and a number of important reef species have been assessed, including fish, corals and turtles. Overall, some 341 coral reef species are threatened,[106] including 200 reef-building corals[107] for which the combined impacts of coral bleaching and disease have been critical drivers of decline, with climate change representing an additional major threat. Staghorn and elkhorn corals were once the two major reef-builders in the Caribbean, but both are now listed as critically endangered. Threatened fish include prime fisheries targets such as larger groupers and bumphead parrotfish, as well as species with restricted ranges[108] for which relatively localized threats may have severe consequences.

MAP 3.4. THREAT TO CORAL REEFS FROM OCEAN ACIDIFICATION IN THE PRESENT, 2030, AND 2050

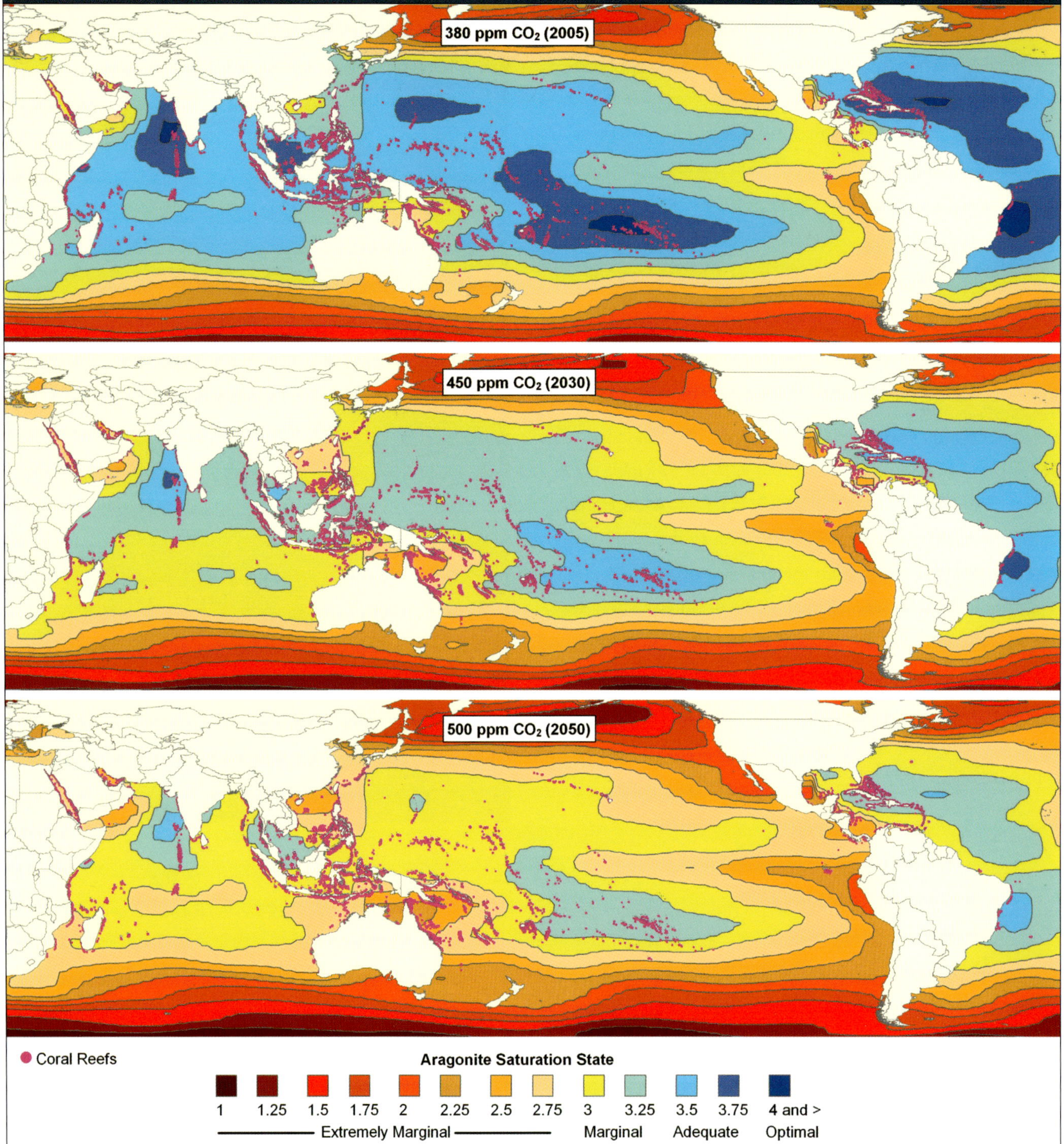

Note: Estimated aragonite saturation state for CO_2 stabilization levels of 380 ppm, 450 ppm, and 500 ppm, which correspond approximately to the years 2005, 2030, and 2050 under the IPCC A1B (business-as-usual) emissions scenario.

Source: Adapted from Cao, L. and K. Caldeira. 2008. "Atmospheric CO_2 Stabilization and Ocean Acidification." *Geophysical Research Letters* 35: L19609.

reduced and reef ecosystems will start losing structural complexity and biodiversity.[94, 126] At CO_2 levels greater than 500 ppm, it is predicted that only a few areas of the world's oceans will be able to support reef-building (calcifying) corals.[126] The current, rapid (geologically speaking) increase in acidification is likely unprecedented in the history of the planet.[94, 127]

Remedies: The slowing and reversal of ocean acidification will ultimately depend on the reduction of CO_2 emissions, perhaps alongside the active removal of CO_2 from the atmosphere, such as through carbon sequestration in soil and vegetation. Many scientists have concluded that 350 ppm is the critical maximum level of atmospheric CO_2 that the world should strive to achieve in order to minimize climate and acidification-related threats to coral reefs and other marine organisms.[94, 128] However, achieving this target depends on the political will of all countries and their agreement to internationally collaborate toward a collective reduction in emissions, as well as concerted effort by people around the world. Little or nothing can be done at the local scale to prevent acidification impacts on reefs, although as with bleaching, it seems possible that multiple stressors acting together may hasten the decline of reefs. Reduction in local pressures may therefore again buy time for the impacts of emission reductions to occur.

SEA LEVEL RISE AND STORMS

To date, climate change has had the most dramatic impact on coral reefs through bleaching events and associated mortality, while the effects of increasing acidification are now becoming detectable. But climate change may also influence reefs in other ways. Sea level rise and high-intensity storms were not explicitly included in the modeling of global-level threats, but represent additional climate-related threats that could impact reefs in the future.

Sea level rise

Global sea level is rising, through both the expansion of water due to warming temperatures and the considerable increase in ocean volumes from the melting of terrestrial ice sheets and mountain glaciers. Together, these changes have already led to an increase in sea level of 20 cm since 1870, with a rate of rise currently at 3.4 mm per year and accelerating.[129, 130] Predictions vary, but by 2100, seas are likely to have risen by 90 to 200 cm over a 1990 baseline level.[129, 131]

Healthy, actively growing reefs are able to "keep up" with rising seas as they build their limestone structures toward the sea surface, and even the more extreme projections point to levels that are probably insufficient to greatly affect reefs in most areas during the period of focus of this work (to 2050). However, the same resilience may not be found in low-lying reef landforms, such as coral islands and atolls, which are the basis for many human settlements, especially in the Pacific. Such islands are formed by sand and coral rock deposited on the reef by waves and currents. For nations like Kiribati, Tuvalu, and the Maldives, made up entirely of coral islands, even small rises in sea level will leave these landforms extremely vulnerable. It is not automatically the case that such islands will erode or be inundated by sea level rise, as the processes by which they were formed will continue, and indeed there is some evidence that under moderate sea level rise some islands may persist or even grow.[132] Even so, it seems that accelerating sea level rise presents a significant threat, and one that is already impacting some islands. The processes of impact may vary: erosion will likely increase,[133] the lowest-lying areas may become inundated during storm events, and rising seas may pollute the shallow freshwater "lens" below the islands, which is critical for drinking water, vegetation, and crops.[134]

Tropical storms

Patterns of tropical storms vary considerably around the world. Equatorial reefs are rarely, if ever, hit by tropical storms, but toward the edges of the tropics, powerful storms form most years. In these areas, individual reefs may be hit multiple times during the same year, or may avoid storm damage for twenty or more years.

Storms can be powerful drivers of change for these coral reefs. They are a natural perturbation in many areas, but nonetheless can dramatically affect reef life by reducing the coral framework to broken rubble that can no longer support high levels of abundance and diversity. Recovery can take years or decades. Where reefs are already weakened by other threats, storms are a complicating factor, bringing an already ailing reef to complete failure.

While it is known that tropical storms exert a powerful influence on reefs, the influence of climate change on storms is less clear.[135] Recent studies have predicted that the frequency of very intense tropical storms may increase as a result of warming sea surface temperatures.[136] Currently, the linkages between climate change and storm activity are still under investigation, and effects will most likely vary regionally.

COMPOUNDING THREATS: DISEASE AND CROWN-OF-THORNS STARFISH

Coral diseases and outbreaks of crown-of-thorns starfish (COTS) can occur naturally on reefs, but are now occurring with increased frequency, often in conjunction with other threats or following coral bleaching events. Disease and COTS (*Acanthaster planci*) were not explicitly included in the modeling because globally consistent data were not available for them, and because uncertainties remain regarding their specific drivers. In the case of disease, its somewhat ambiguous role as both a threat and symptom of other threats represented a further obstacle to modeling its impacts on reefs, while for COTS, at least some of the proposed drivers (such as overfishing, terrestrial runoff) are already included in the model. We describe these key threats below and discuss their co-occurrence with the modeled threats.

DISEASE

Diseases are a natural feature in any ecosystem and are present in background populations of most species. Both in terms of prevalence and geographic distribution, coral diseases have increased in recent years.[137] The drivers of increasing disease occurrence are still not clearly understood, but it is probable that corals have become more susceptible to disease as a result of degraded water quality and that warming due to climate change may cause some pathogens to become more virulent and may also affect a coral's immunodefense capabilities.[138] There is strong evidence that disease outbreaks have followed coral bleaching events.[139]

Undoubtedly, disease has already altered reef systems in the Caribbean.[140] White-band disease has virtually wiped out elkhorn (*Acropora palmata*) and staghorn (*Acropora cervicornis*) corals, which were once the two greatest reef-builders in the region. [141] Another disease, which affects the long-spined sea urchin (*Diadema antillarum*), has also dramatically altered Caribbean reefs.[142] These urchins are major grazers of algae on reefs, particularly in areas where overfishing has removed most grazing fish. An outbreak of an unknown disease among urchins in 1983–84 was followed by a surge in algal growth on corals in the absence of these grazers. In recent years, urchins have recovered in some parts of the Caribbean, such as along the north coast of Jamaica, with associated reductions in algae and some regeneration of corals.[143]

MAP 3.5 GLOBAL INCIDENCE OF CORAL DISEASE, 1970–2010

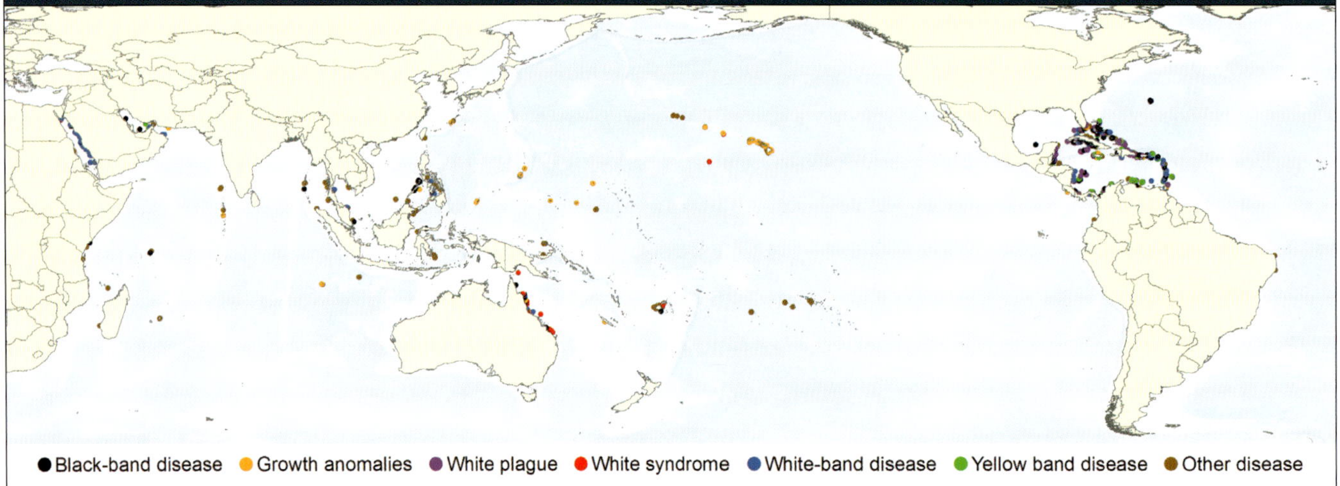

Note: This map provides an indication of the broad patterns of coral disease, but is incomplete because many coral reef locations are unexplored, and not all observations of coral disease are reported. "Other" includes skeletal eroding band, brown band, atramentous necrosis, trematodiasis, ulcerative white spots, and other syndromes that are poorly described.
Source: ReefBase Coral Disease data set and UNEP-WCMC Global Coral Disease database, observations of coral disease 1970–2010.

BOX 3.10 REEF STORY

Brazil: Coral Diseases Endanger Reefs

Brazil's Abrolhos Bank contains some of the largest and richest coral reefs in the South Atlantic. In the last 20 years, the area's coastline has experienced increased tourism, urbanization, and large-scale agriculture, leading to discharge of untreated waste and contamination of the region's reefs. As a result, the prevalence of coral disease has dramatically escalated off the Brazilian coastline in recent years. Furthermore, studies have linked the global proliferation of coral diseases to elevated seawater temperature, suggesting that climate change will lead to even greater incidences of disease in Brazil in the future. If the area's corals continue to die off at the current rate, Brazil's reefs will suffer a massive coral cover decline in the next 50 years. *See full story online at www.wri.org/reefs/stories.*

Story provided by Ronaldo Francini-Filho and Fabiano Thompson of the Universidade Federal da Paraiba and Rodrigo Moura of Conservation International, Brazil.

Coral disease research is still in its infancy, but due to the urgency of the problem, research is currently being undertaken hand-in-hand with management efforts. Current efforts to address the threat of coral disease are aimed at understanding its drivers and impacts and how these may be affected by climate change. One important part of this work involves compiling both baseline and long-term data about the distribution and prevalence of coral disease, in order to examine spatial and temporal patterns and trends and to identify factors that influence vulnerability and resilience.[144] Map 3.5 provides an indication of the broad patterns of disease, but this map shows only a fraction of disease incidence due to limitations in reporting. Given that diseases are often more problematic where corals are already under stress, management efforts such as protecting water quality, preserving functional diversity, and reducing other threats to reefs may help to lessen the occurrence and impacts of disease.[145]

CROWN-OF-THORNS STARFISH (COTS)

Another natural threat with severe consequences for reefs is the occurrence of plagues or outbreaks of the crown-of-thorns starfish (COTS) across the Indo-Pacific region.[146] These starfish are natural predators of coral, and usually occur at low densities on reefs. However, if their numbers reach outbreak proportions, they can kill vast stretches of coral, having an impact similar to that of an extreme coral bleaching event. Since the 1950s, such outbreaks have been recorded across much of the Indo-Pacific, and areas of recent outbreaks include reefs in the Red Sea, East Africa, East and Southeast Asia, and the Pacific.[47] The exact cause of these outbreaks remains unclear. Some occurrences may simply be natural fluctuations in population size, but there are indications that overfishing of predatory fish, such as wrasses and triggerfish, may play a part.[147] Nutrient pollution of coastal waters and estuaries may also contribute to outbreaks by stimulating the growth of algae, the preferred food for COTS larvae.[148]

In a few places, efforts to physically remove COTS from relatively confined reef areas (such as around small islands or adjacent to tourist areas) have been successful. Larger-scale control programs have also been attempted, most notably in the Ryukyu Islands of Japan, but such efforts are now generally regarded as impossible. The best hope for reducing further outbreaks or minimizing their impact on reefs is likely to come from combating specific threats that cause outbreaks (such as overfishing and terrestrial runoff of nutrients).

The following chapter provides a summary of results of the *Reefs at Risk* modeling of current and future threats to the world's coral reefs.

Chapter 4. REEFS AT RISK: RESULTS

Our analysis indicates that more than 60 percent of the world's coral reefs are under immediate and direct threat from one or more of the combined local pressures of overfishing and destructive fishing, coastal development, watershed-based pollution, and marine-based pollution and damage. When recent thermal stress is factored in, the overall measure of present threat rises to 75 percent of all reefs. High and very high threats occur on almost 30 percent of all reefs, 40 percent when recent thermal stress is included. Overfishing is by far the most widespread local threat, affecting more than half of all reefs, while coastal development and watershed-based pollution each affect about 25 percent. The region most affected by local threats is Southeast Asia, where almost 95 percent of reefs are threatened.

These results are presented in more detail in this chapter. Detailed regional findings and underlying drivers and responses are presented in Chapter 5.

In order to gauge ongoing trends in risks to reefs, we also conducted a separate analysis that considers changes in human pressure since 1998. This involved re-applying the less refined modeling approach from 1998 to more current data on population and development (Box 4.1).

PRESENT LOCAL THREAT BY TYPE

The following sections present the results of the analysis of individual threats, as well as the threats combined into the integrated threat index. The individual threat indicators are designed to identify areas where, in the absence of good management, coral reef degradation is probably occurring or where current levels of human activity suggest that it is likely to happen in the near future. Definitions of low, medium, and high categories for each threat are available in the Technical Notes at www.wri.org/reefs. In the following descriptions, the term "threatened" refers to reefs at medium or higher threat.

Coastal development

Development along the coast threatens almost 25 percent of the world's reefs, of which more than 10 percent face a high threat. The largest proportion of threatened reefs are in Southeast Asia, where small islands with densely populated coastlines put pressure on at least one-third of the region's corals. Coastal development is also a major threat in the Indian Ocean and the Atlantic; more than a quarter of reefs are threatened in each region. These figures are likely to be conservative, due to the inability of available data to capture very recent development.

FIGURE 4.1. REEFS AT RISK FROM COASTAL DEVELOPMENT

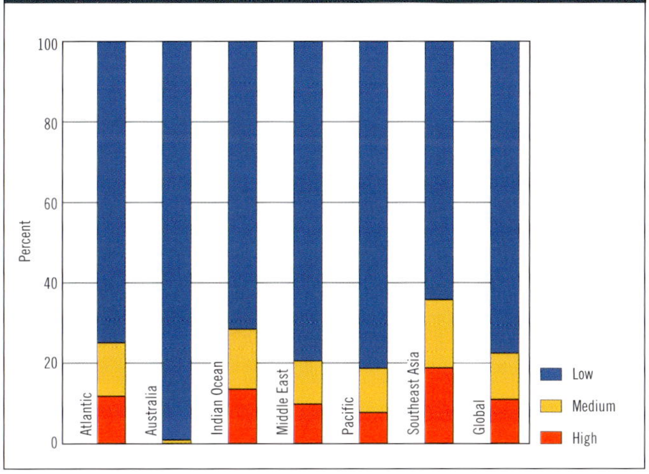

Watershed-based pollution

More than one-quarter of the world's reefs are threatened by watershed-based pollution (including nutrient fertilizers, sediment, pesticides, and other polluted runoff from the land), with about 10 percent considered to be highly threatened. Southeast Asia surpasses all other regions with 45 percent of reefs threatened. The magnitude of threat in this region is driven by a high proportion of agricultural land use, steep terrain, heavy precipitation, and close proximity of reefs to land. More than 30 percent of reefs in the Indian Ocean region are similarly threatened by watershed-based pollution. The majority of the threat in this region originates from heavily cultivated areas of coastal East Africa and Madagascar. About 5 percent of reefs in Australia were rated as threatened by watershed-based pollution, mostly the nearshore reefs in the southern Great Barrier Reef. This indicator focuses on erosion and sediment dispersion in river plumes, so it likely underestimates the threat from nutrients and pesticides, which tend to travel further from river mouths.

Marine-based pollution and damage

Marine-based sources of pollution and damage threaten approximately 10 percent of reefs globally, with only about 1 percent at high threat. This pressure is widely dispersed around the globe, emanating from ports and widely distributed shipping lanes. The Atlantic, Middle East, and Australia are the regions most affected. The threat in the Atlantic is mainly influenced by the large number of commercial and cruise ship ports, and associated shipping traffic. Middle Eastern reefs are affected by a vast number of offshore oil rigs. Australia has relatively few large ports, but important shipping lanes pass inside and across the Great Barrier Reef, although in reality these are relatively well-managed and represent a potential threat that to date has only had a minimal impact. Despite advances in monitoring shipping traffic, these data are incomplete in that they exclude all fishing and smaller recreational vessels, so this estimate should be considered conservative.

FIGURE 4.2. REEFS AT RISK FROM WATERSHED-BASED POLLUTION

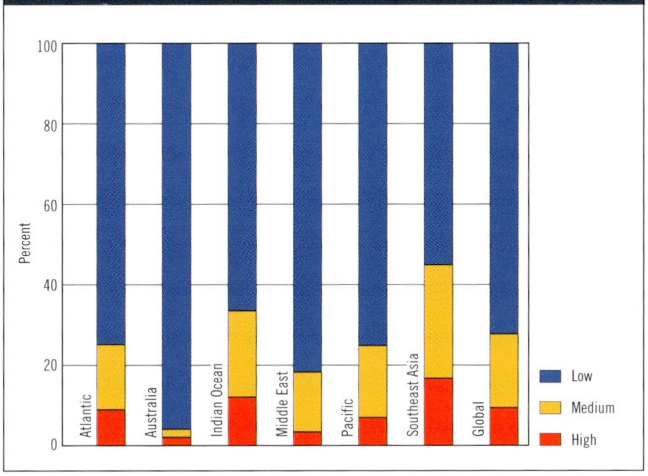

FIGURE 4.3. REEFS AT RISK FROM MARINE-BASED POLLUTION AND DAMAGE

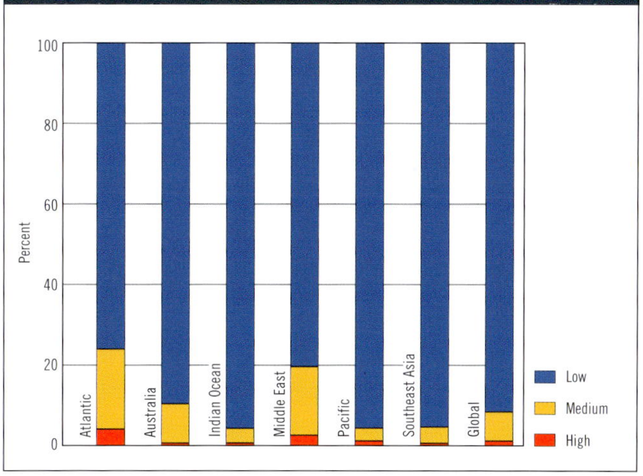

Overfishing and destructive fishing

Unsustainable fishing is the most pervasive of all local threats to coral reefs. More than 55 percent of the world's reefs are threatened by overfishing and/or destructive fishing, with nearly 30 percent considered highly threatened. Reefs in Southeast Asia are most at risk, with almost 95 percent of reefs affected. Densely populated coastlines, shallow and easily accessible fishing grounds, as well as the highest global occurrences of blast and poison fishing contribute to the threat in this region. Reefs in the Indian Ocean and the Atlantic are also significantly threatened by overfishing, with nearly 65 percent and 70 percent of reefs affected, respectively. The model is conservative for many remote coral reefs since it focuses on the fishing practices of populations living adjacent to reefs. In reality, even many of the most remote coral reefs are now heavily fished, often illegally, for valuable "target species" such as sharks.

FIGURE 4.4. REEFS AT RISK FROM OVERFISHING AND DESTRUCTIVE FISHING

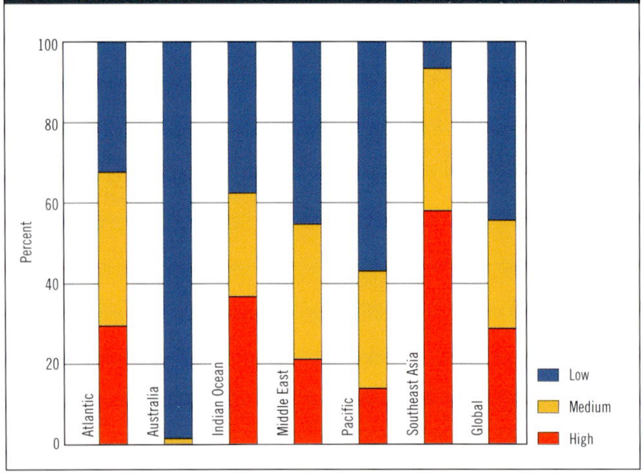

Past thermal stress

Mapping of past thermal stress on coral reefs (1998–2007) suggests that almost 40 percent of reefs may have been affected by thermal stress, meaning they are located in areas where water temperatures have been warm enough to cause severe bleaching on at least one occasion since 1998. Map 3.2 presents the locations of satellite-detected thermal stress and coral bleaching observations, and Figure 4.5 shows the

estimated percentage of reefs per region that experienced thermal stress based on this map. These satellite-derived estimates are a good approximation of threat, but do not include the possible influence of past temperature variability, which may produce a degree of acclimation or adaptation in reefs that have been subjected to past thermal stress.[149, 150] Conversely, reefs in areas with little historic variation in temperatures may be more vulnerable to bleaching from even very small deviations in temperature.[151] At local scales, reefs may be further buffered from temperature stress by small-scale influences such as shading, currents, strong tidal flows, and cold-water upwelling.[152] In the *Reefs at Risk Revisited* model, past thermal stress is treated as an additional threat acting upon the integrated local threats.

FIGURE 4.5. THERMAL STRESS ON CORAL REEFS BETWEEN 1998 AND 2007

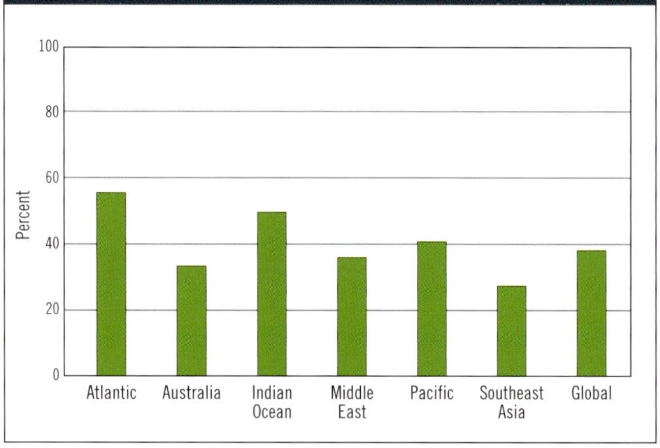

PRESENT INTEGRATED THREATS TO CORAL REEFS

The four local threats to coral reefs described in this chapter are combined in the integrated local threat index. This index is presented on the world map of coral reefs inside the front cover. Figure 4.6 provides a summary of the four individual local threats and the integrated local threat index. The sixth column reflects the full integrated threats to the world's reefs and incorporates past thermal stress.

Globally, more than 60 percent of the world's coral reefs (about 150,000 sq km of reef) are threatened by local activities and 75 percent are threatened when past thermal stress is included. Figure 4.7 presents the integrated threat results according to the amount of reef area threatened per region. A more detailed description is presented in Chapter 5.

Table 4.1 provides a summary of the integrated threats to coral reefs by region and for the 15 countries and territories with the most coral reefs.

PHOTO: ARC CENTRE OF EXCELLENCE FOR CORAL REEF STUDIES

FIGURE 4.6. REEFS AT RISK FROM INDIVIDUAL LOCAL THREATS AND ALL THREATS INTEGRATED

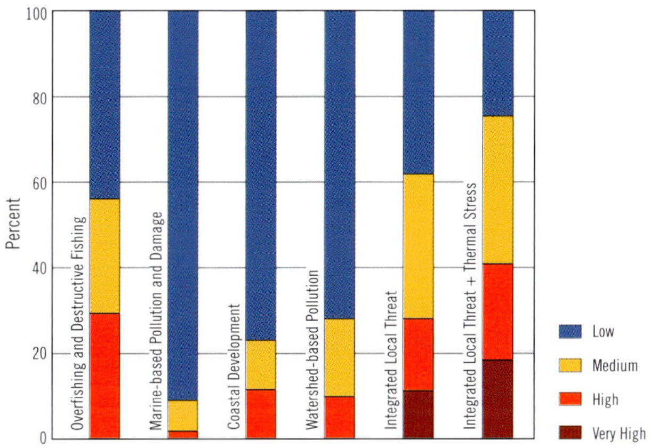

Note: The first four columns reflect individual, local threats to the world's coral reefs. The fifth column (integrated local threat) reflects the four local threats combined, while the sixth column also includes past thermal stress.

FIGURE 4. 7. REEFS AT RISK FROM INTEGRATED LOCAL THREATS (BY AREA OF REEF)

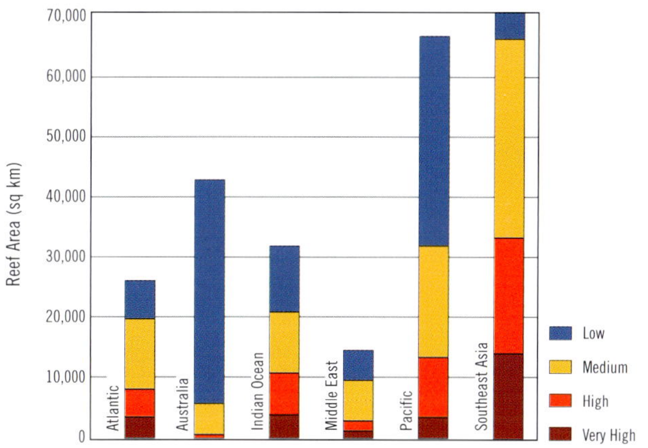

Note: Amount of reef area (in sq km) in each region classified by integrated local threat. Further details, including information on past thermal stress, can be seen in the regional breakdowns in Chapter 5.

TABLE 4.1 INTEGRATED THREAT TO CORAL REEFS BY REGION AND COUNTRIES/TERRITORIES WITH THE HIGHEST CORAL REEF AREA

			Integrated Local Threats								
Region	Reef Area (sq km)	Reef area as percent of global	Low (%)	Medium (%)	High (%)	Very High (%)	Threatened (medium or higher) (%)	Severe thermal stress (1998–2007) (%)	Integrated Local + Thermal Stress Threatened (medium or higher) (%)	Coastal Population (within 30 km of reef)[b] '000	Reef Area in MPAs (%)
Atlantic	25,849	10	25	44	18	13	75	56	92	42,541	30
Australia [a]	42,315	17	86	13	1	<1	14	33	40	3,509	75
Indian Ocean	31,543	13	34	32	21	13	66	50	82	65,152	19
Middle East	14,399	6	35	44	13	8	65	36	76	19,041	12
Pacific	65,972	26	52	28	15	5	48	41	65	7,487	13
Southeast Asia	69,637	28	6	47	28	20	94	27	95	138,156	17
Global	**249,713**	**100**	**39**	**34**	**17**	**10**	**61**	**38**	**75**	**275,886**	**27**
Key Countries											
Australia [a]	41,942	17	86	13	1	<1	14	33	40	3,507	75
Indonesia	39,538	16	9	53	25	12	91	16	92	59,784	25
Philippines	22,484	9	2	30	34	34	98	47	99	41,283	7
Papua New Guinea	14,535	6	45	26	22	7	55	54	78	1,570	4
New Caledonia	7,450	3	63	30	6	<1	37	39	57	210	2
Solomon Islands	6,743	3	29	42	24	6	71	36	82	540	6
Fiji	6,704	3	34	34	21	10	66	54	80	690	32
French Polynesia	5,981	2	76	15	7	2	24	13	33	269	1
Maldives	5,281	2	62	33	4	1	38	74	87	357	<1
Saudi Arabia	5,273	2	39	44	11	6	61	47	73	7,223	1
Federated States of Micronesia	4,925	2	70	23	6	1	30	31	52	100	<1
Cuba	4,920	2	5	71	14	10	95	36	97	4,430	14
Bahamas	4,081	2	40	52	6	2	60	47	79	303	3
Madagascar	3,934	2	13	35	34	18	87	41	94	2,235	2
Hawaii (US)	3,834	2	83	3	6	9	17	11	28	1,209	85

Notes: a. The Australia region includes the Australian territories of Christmas Island and Cocos/Keeling Islands, whereas Australia in "Key Countries" does not.
b. Population statistics represent the human population, both within 10 km of the coast as well as within 30 km of a coral reef.

Sources:

1. **Reef area estimates:** Calculated at WRI based on 500-m resolution gridded data assembled under the *Reefs at Risk Revisited* project from Institute for Marine Remote Sensing, University of South Florida (IMaRS/USF), Institut de Recherche pour le Développement (IRD), UNEP-WCMC, The World Fish Center, and WRI (2011).
2. **Coastal population within 30 km of reef:** Derived at WRI from LandScan population data (2007) and World Vector Shoreline (2004).
3. **Number of MPAs:** Compiled at WRI from the World Database of Protected Areas (WDPA), ReefBase Pacific, The Nature Conservancy, and the Great Barrier Reef Marine Park Authority.

BOX 4.1. TEN YEARS OF CHANGE: 1998 TO 2007

Much has changed—for better and worse—since the first *Reefs at Risk* was released in 1998. Human pressure on coral reefs has increased significantly, management of coastal ecosystems has improved in some areas, and our ability to estimate threats to reefs has improved with advances in data from satellites, new maps, and new modeling methods. For example, the global map of coral reefs compiled for this analysis has a resolution of 500m, which is 64 times more detailed than the map used in the 1998 *Reefs at Risk* analysis.

Given the considerable improvements in input data and modeling methods, it is not possible to make a direct comparison of the findings highlighted in this report with those published in 1998. The change in the reef map alone alters results significantly; many additional reefs have been mapped, notably in relatively remote, low pressure areas. But in order to evaluate change in pressure on coral reefs since 1998, we undertook a separate comparative analysis, with results presented below.

Analysis Approach: It was not possible to examine changes in reef threat since 1998 by using the latest model rules with 1998 data, because many of the newly included data sets—such as hotel locations—do not exist for 1998. Therefore, we evaluated these changes through a two-part analysis. First, we re-ran our analyses from 1998, using the new (500-m resolution) coral reef map with the 1998 threat data, which allowed us to compensate for the improved resolution of the new map. We refer to these results as "1998 revised." (The global percentage of threatened reefs dropped from 58 percent in the original 1998 analysis, to 54 percent under the "1998 revised" analysis, reflecting the fact that the updated reef map now includes many remote reefs which are under relatively low threat.) Second, we used the modeling method from 1998 with current threat data sets (for example, population in 2007) to develop indicators of threat in 2007 (referred to as "ten year update"), which can be compared to the "1998 revised" results. Results of this comparison are described below and summarized in Figures 4.8 and 4.9. The comparison does not include threats from thermal stress or changing ocean chemistry, as these were not included in the 1998 analysis.

Note: In undertaking this work it became apparent that the 1998 models gave higher predictions of threat than the models used in the rest of this report. We emphasize, though, that this should not be interpreted as a reduction in overall threat levels to coral reefs, but rather a reflection of more accurate models combined with the improved global reef maps, which include better coverage of remote reefs far from human threats.

Results: The percentage of the world's reefs rated as threatened by integrated local threats (medium threat or higher) increased by more than 30 percent between 1998 and 2007 (from 54 percent to 70 percent.)

Ten Years of Change by Threat

By far the greatest driver of increased pressure on reefs is overfishing and destructive fishing, which has increased by roughly 80 percent since 1998. The greatest increase in overfishing was observed in the Pacific, where previously this was a minor threat. Large increases in overfishing also occurred in the Indian Ocean, Middle East, and Southeast Asia. This change is driven largely by coastal population growth near reefs.

continued

MAP 4.1. CHANGE IN LOCAL THREAT BETWEEN 1998 AND 2007

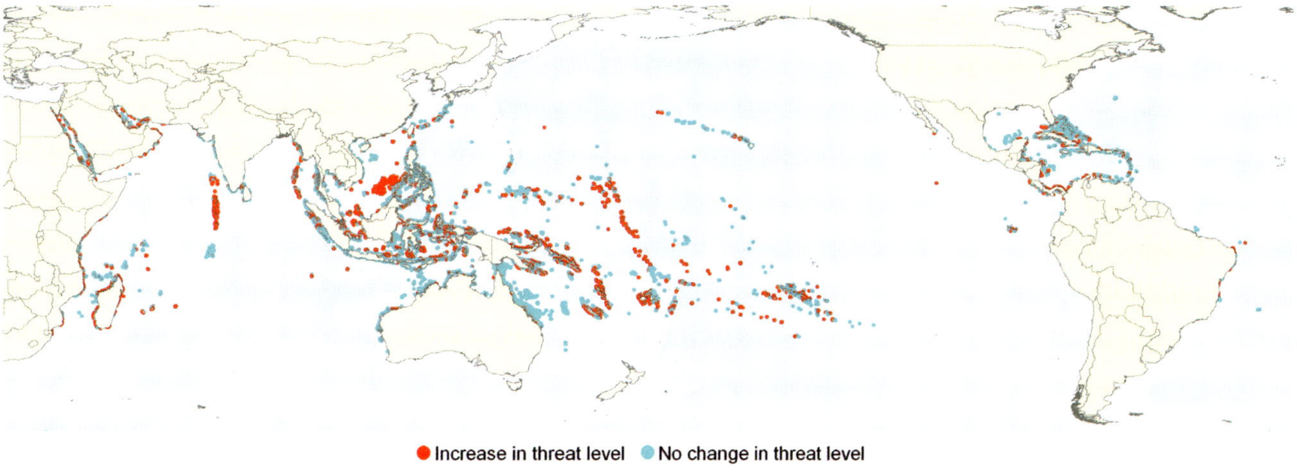

Note: These results use the 1998 modeling methodology and new coral reef data.

BOX 4.1. *continued*

Pressure on reefs from coastal development and watershed-based pollution has also increased significantly—both by about 15 percent over 1998 levels. Coastal population growth, increased runoff, sewage discharge, and conversion of coastal habitats all increase sediment and pollutants reaching reefs. Marine-based pollution and damage increased by a similar proportion, with the largest increase in pressure in the Atlantic.

FIGURE 4.8. REEFS AT RISK BY THREAT IN 1998 AND 2007 (PERCENT AT MEDIUM OR HIGH THREAT)

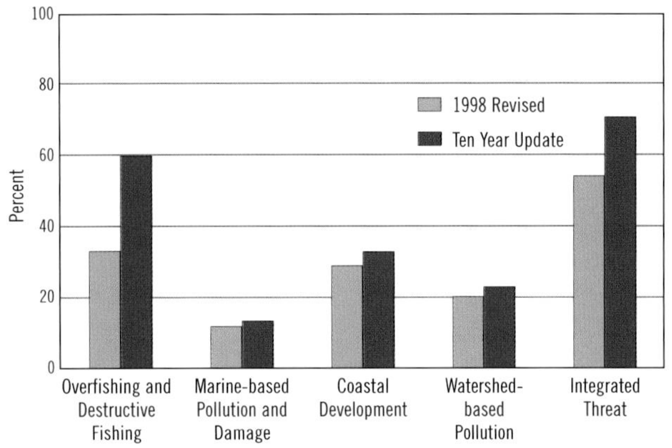

Note: Percent of the world's reefs threatened by local activities in 1998 and 2007. These results use the 1998 modeling methodology and new coral reef data.

Ten Years of Change by Region

Local threats to coral reefs have increased in all regions, but most notably in the Pacific and Indian Ocean. In the Pacific, the proportion of threatened reefs rose by about 60 percent. This increase was driven mostly by increased overfishing pressure, although watershed-based pollution and coastal development have also increased in many areas.

In the Indian Ocean, the percentage of threatened reefs has increased by over 40 percent; the largest driver is population growth, which in turn drives overfishing pressure.

In the Middle East, the percent of reefs rated as threatened increased by only 10 percent over the ten years, but the proportion of highly threatened reefs rose markedly, driven by increases in overfishing, marine-based pollution and damage, and coastal development.

In Southeast Asia, the local threat to reefs increased by about 20 percent, although the threat in this region was already very high in 1998. Overfishing pressure in many areas shifted from medium to high threat, and coastal development pressure increased in many areas.

In the Atlantic region, the local threat increased by nearly 20 percent. Many new areas are now threatened by watershed-based pollution or marine-based pollution and damage. In addition, many areas shifted from medium to high overfishing threat.

Australia had the lowest apparent increase in local pressure on reefs over the ten-year period, with a slight increase in both the percentage of reefs classified as threatened, and the proportion at high threat. Of the four local threats, watershed-based pollution showed the greatest increase.

FIGURE 4.9. REEFS AT RISK FROM INTEGRATED LOCAL THREATS IN 1998 AND 2007

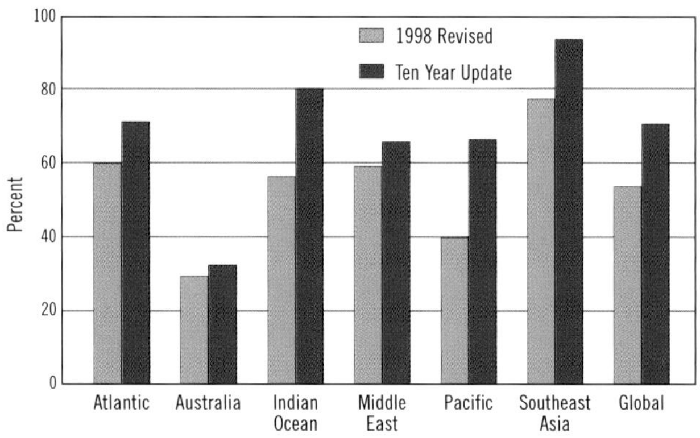

Note: Percent of reefs threatened by integrated local threat per region in 1998 and 2007. These results use the 1998 modeling methodology and new coral reef data.

FUTURE INTEGRATED THREATS TO CORAL REEFS

In this section, we look ahead to the likely state of the world's reefs over the next 20 to 40 years. First, we briefly outline the main drivers of change, and then present modeling forecasts for 2030 and 2050.

Key drivers of reef condition

At present, local human activities, coupled with past thermal stress, threaten an estimated 75 percent of the world's reefs. Without intervention, these pressures have trajectories slated to escalate into the future. Global human population is projected to reach 8.9 billion by 2050 – a 35-percent increase over 2007 levels[153] with much of the growth in coral reef and other developing nations, increasing pressure on reefs.

The single greatest growing threat to coral reefs is the rapid increase in greenhouse gases in the atmosphere, including carbon dioxide (CO_2), methane, nitrous oxide, and halocarbons. Since preindustrial times, atmospheric concentrations of all of these gases have increased significantly, and in the case of CO_2, which contributes the most to both warming and acidification, concentrations have risen by over 35 percent.[130] During the past 10 years, almost 40 percent of coral reefs have experienced thermal stress at a level sufficient to induce severe coral bleaching (Figure 4.5). Under a "business-as-usual" scenario, our models suggest that roughly 50 percent of the world's reefs will experience thermal stress sufficient to induce severe bleaching in five out of ten years during the 2030s. During the 2050s, this percentage is expected to grow to more than 95 percent (Map 3.3). Most evidence suggests that these extreme levels of coral bleaching will likely lead to the degradation of coral reefs worldwide.

In addition, increasing CO_2 emissions are dissolving into the oceans and changing the chemical composition of seawater. Increased CO_2 elevates the acidity of seawater and reduces the saturation state of aragonite, the mineral corals use to build their skeletons. The best available modeling suggests that by 2030, fewer than half of the world's reefs will be in areas where aragonite levels are adequate for coral growth; that is, where the aragonite saturation state is more than 2.75. By 2050, only about 15 percent of reefs will be in areas where aragonite levels are adequate for growth (Map 3.4).

There are, of course, uncertainties associated with these predictions. Future warming projections rely on assumptions about future greenhouse gas emissions, modeled estimates of atmospheric and ocean warming, and thresholds for prompting damaging coral bleaching. Many factors influence coral bleaching, at multiple scales, which are not yet fully understood; past thermal stress has not always been sufficient to predict occurrence, severity, or mortality from bleaching, but it remains our best indicator. The prediction of future threat posed by acidifying seas relies on scenarios of future CO_2 emissions, models of ocean chemistry, and the best current scientific understanding of the critical importance of aragonite for coral growth. There is also some evidence that coral species may vary in their ability to deal with increased acidity[154] and that physical or physiological mechanisms may help to reduce the effects of acidification. However, the projections for future acidification are so high that these factors will have little or no overall long-term impact. It is also important to note that these projections assume that current local threats remain constant in the future, and do not account for potential changes in human pressure, management, or policy, which could influence overall threat ratings.

Threat in 2030

By the 2030s, our estimates predict that more than 90 percent of the world's reefs will be threatened by local human activities, warming, and acidification, with nearly 60 percent facing high, very high, or critical threat levels. Thirty percent of reefs will shift from low threat to medium or higher threat specifically due to climate or ocean chemistry changes. An additional 45 percent of reefs that were already impacted by local threats will shift to a higher threat level by the 2030s due to climate or ocean chemistry changes (Figure 4.10). Thermal stress will play a larger role in elevating threat levels than acidification by 2030, though about half of all reefs will be threatened by both conditions. As shown in Figure 4.10, the predictions for thermal stress and acidification in the 2030s have the most dramatic effect on the reefs in Australia and the Pacific, pushing many reefs from low to threatened categories. In addition, climate-related threats in parts of Southeast Asia will compound already high local threat levels in that region.

Maps 4.2a and 4.2b show reefs classified by estimated threat level today and in 2030. By the 2030s, the predicted increase in threat due to warming and acidification is apparent across all regions of the world. Many of Australia's reefs will shift from low to medium or high threat. This is also true in Papua New Guinea and much of the western Pacific. Increases in threat are also apparent for many islands in the Indian Ocean and for much of the Caribbean coast of Central America (see Chapter 5). However, in 2030, there will still be some reefs under low threat in all regions of the world, including parts of the Bahamas in the Caribbean; French Polynesia and the Northwest Hawaiian Islands in the

Pacific; the Red Sea in the Middle East; the Maldives, Seychelles and Mauritius in the Indian Ocean; the southern Great Barrier Reef in Australia, and a few reefs in Central Indonesia in Southeast Asia.

Threat in 2050

By the 2050s, estimates predict that almost no reefs will be under low threat and only about one-quarter will be under medium threat, with the remaining 75 percent at a high, very high, or critical threat levels (Figure 4.10). Looking only at global threats, high thermal stress will be ubiquitous by 2050, and more than 20 percent of reefs are projected to be at high risk for both thermal and acidification threats. As shown in Map 4.2c, by 2050, the few large expanses of reefs rated as low threat in 2030 will have become threatened, with most of these areas under medium threat. Many other reefs are projected to increase from medium to higher threat ratings. A few small areas of reef are projected to remain under low threat in Australia and the south Pacific.

The maps and summary statistics presented in this chapter incorporate current local threats and future global-level threats. If future population growth, coastal development, and agricultural expansion were considered, the projections of the threat to reefs would be even higher. It is important to remember that the results presented here are projections and are not foregone conclusions. This analysis highlights the urgent need for global action to curtail greenhouse gas emissions, in parallel with local actions to lessen the immediate pressures on coral reefs. Controlling local threats to coral reefs will be critical to ensuring their survival in the face of heavy human pressure in coastal regions, and growing threats from climate change and ocean acidification.

FIGURE 4.10 REEFS AT RISK PROJECTIONS: PRESENT, 2030, AND 2050

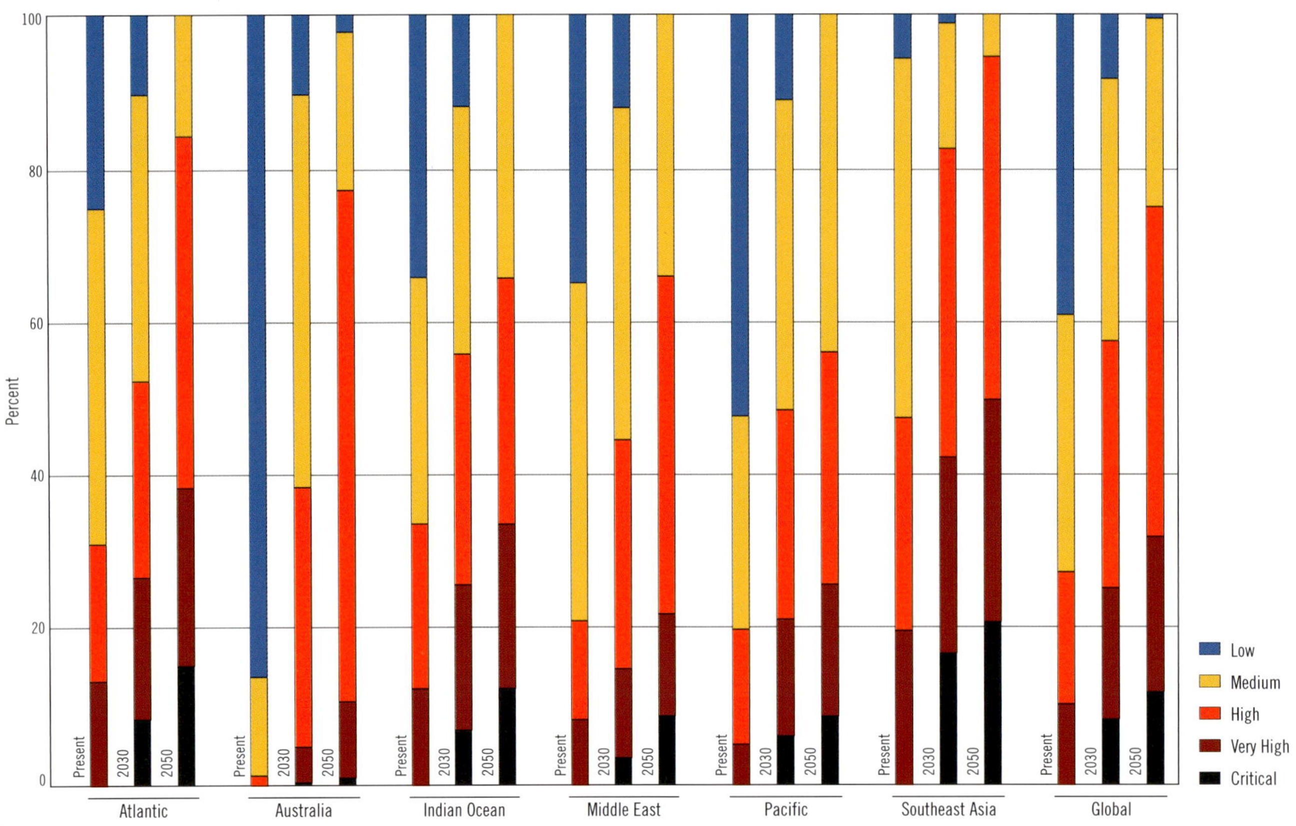

Note: "Present" represents the Reefs at Risk integrated local threat index, without past thermal stress considered. Estimated threats in 2030 and 2050 use the present local threat index as a base and also include projections of future thermal stress and ocean acidification. The 2030 and 2050 projections assume that current local threats remain constant in the future, and do not account for potential changes in human pressure, management, or policy, which could influence overall threat ratings.

MAP 4.2. a, b, and c. REEFS AT RISK IN THE PRESENT, 2030, AND 2050

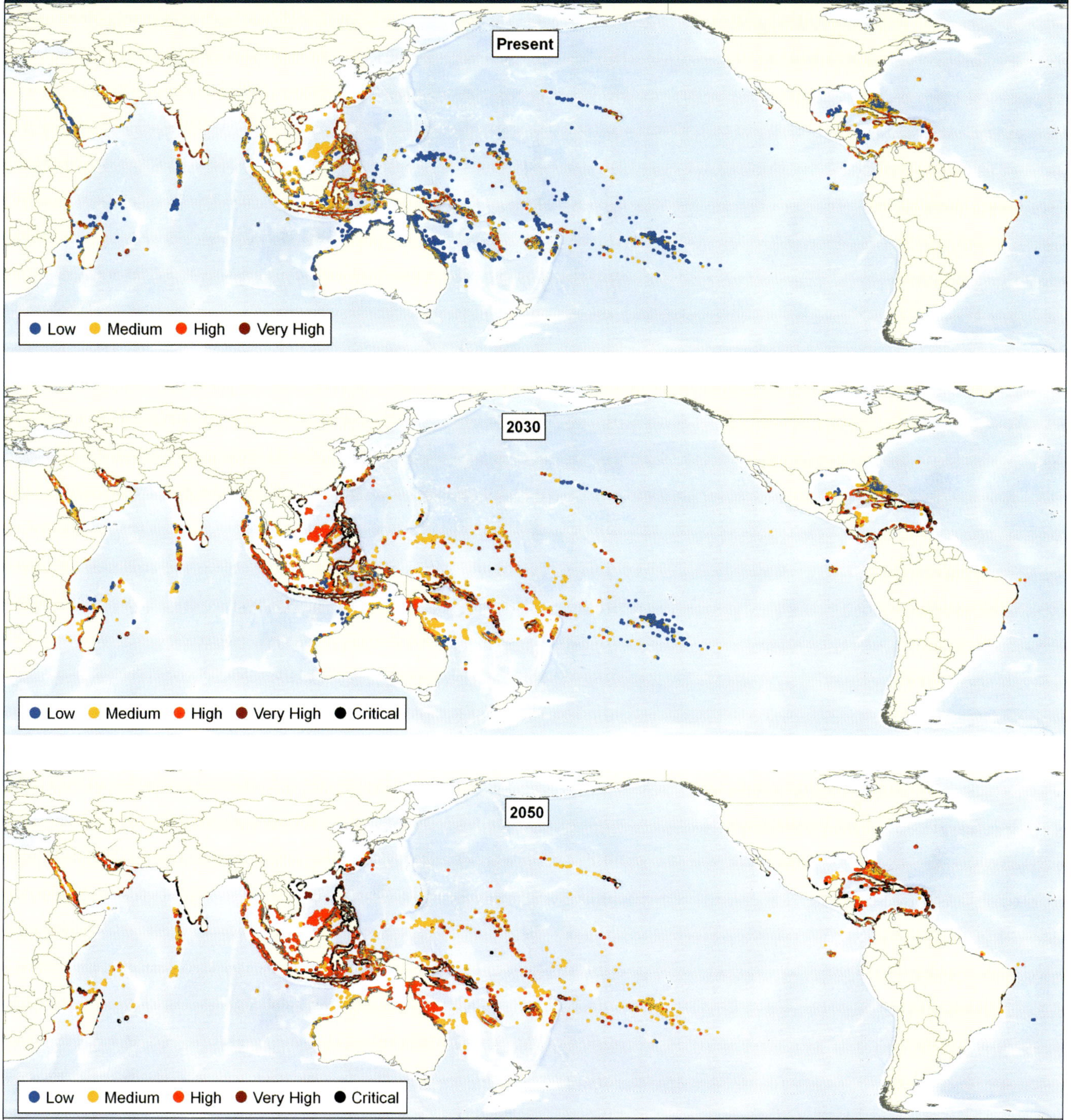

Note: Map 4.2a shows reefs classified by present integrated threats from local activities. Maps 4.2b and 4.2c show reefs classified by integrated local threats combined with projections of thermal stress and ocean acidification for 2030 and 2050, respectively. Reefs are assigned their threat category from the integrated local threat index as a starting point. Threat is raised one level if reefs are at high threat from either thermal stress or ocean acidification, or if they are at medium threat for both. If reefs are at high threat for both thermal stress and acidification, the threat classification is increased by two levels. The analysis assumes no increase in future local pressure on reefs, and no reduction in local threats due to improvements in management.

Chapter 5. REGIONAL SUMMARIES

PHOTO: DAVID BURDICK

At a global scale, the threats facing the world's coral reefs present a considerable challenge to human society. However, it is only by understanding the root causes and impacts of these threats in specific locations that we can begin to develop coherent responses. The key drivers of threats, the current condition and future risk to reefs, and the management measures being utilized to protect reefs are highly variable from place to place. This chapter explores reef distribution, status, threats, and management responses in each of six major coral reef regions.

MIDDLE EAST

The region. The seas surrounding the Arabian Peninsula—Red Sea, Gulf of Aden, Persian (or Arabian) Gulf, Gulf of Oman, and Arabian Sea—represent a distinct coral reef region in the western Indian Ocean, separated by wide areas devoid of reefs along the coastlines of Somalia and Pakistan. This small region has about 6 percent of the world's coral reefs (about 14,000 sq km), almost all of which are found on the continental margins in fringing, barrier, and platform reefs. There are virtually no perennial rivers, so terrestrial sediments only flow into adjacent waters during rare flooding. The Red Sea and Gulf of Aden have narrow shelves, with deep waters nearby. In contrast, the Persian Gulf is shallow, averaging only 35 m deep. This Gulf only formed as sea levels rose after the last ice age. It is subject to wide temperature fluctuations and high salinities, linked to high evaporation and the lack of freshwater input.

Biodiversity. The Red Sea has a rich reef fauna, including many endemic species found nowhere else on earth. For example, about 14 percent of Red Sea reef fish are endemic, including seven unique species of butterfly fish.[155] The Gulf of Aden, including the island of Socotra, has few reefs. This area shares most species with the Red Sea, but also has a number of unique reefs that are seasonally colonized by large kelps (seaweeds) during cold periods of nutrient-rich upwellings in the summer months. By contrast, the Persian Gulf has very low diversity, although many species are uniquely adapted to the harsh conditions of temperature and salinity. These species include corals that are better able to survive in both cool winter temperatures and much warmer summer temperatures than on any other coral reef, providing a living laboratory for better understanding the effects of temperature and the potential for adaptation.

People and reefs. The Middle East has some of the largest tracts of sparsely inhabited continental coastlines in the world, but also has some of the fastest growing populations, enhanced by immigration and tourism. Coastal devel-

MAP 5.1. REEFS AT RISK IN THE MIDDLE EAST

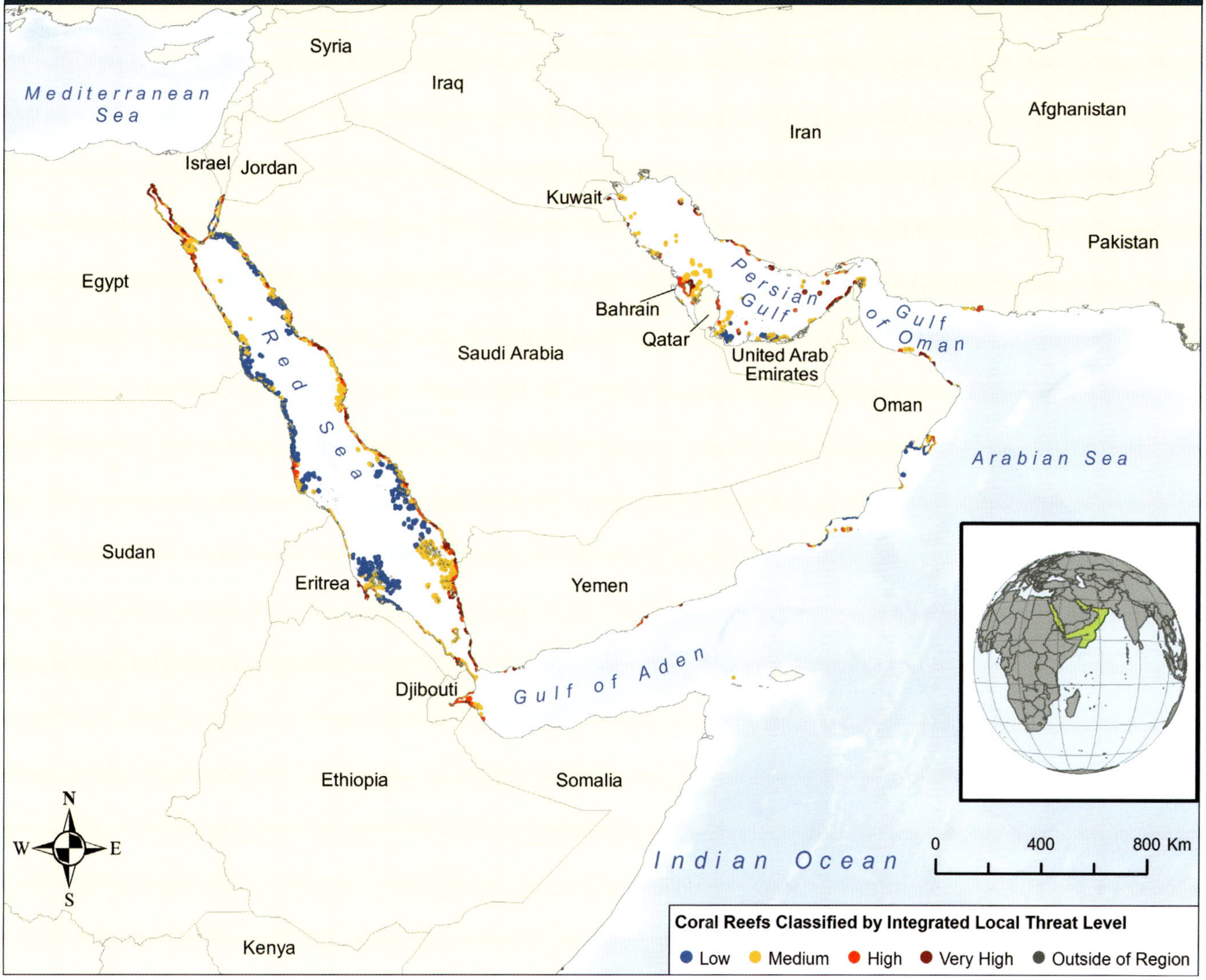

opment, particularly in some of the very wealthy economies, is bringing profound changes in a few areas such as the southern Persian Gulf and the Red Sea port of Jeddah. In these places, extensive areas of shallow water have been filled in for industrial, urban, and tourism infrastructure, resulting in direct impacts (such as loss of habitat and ecosystems) as well as indirect impacts (such as alteration to sediment transport and current patterns) over wide areas. The Persian Gulf also has the world's largest oil reserves, with widespread development of oil rigs and pipelines, coastal storage and refining facilities, and busy shipping lanes. Between 20 and 40 percent of the world's oil supply passes from the Gulf through the Strait of Hormuz each year.[156]

In this region, about 19 million people live on the coast within 30 km of a coral reef.[157] Fishing remains widespread, and is particularly important in the non-oil-producing nations. Fishing in Yemen and Oman mainly takes place in the highly productive waters associated with offshore upwelling and not on the reefs. Tourism is relatively small-scale in many areas, but Egypt and Jordan have important coral reef-related tourism.

Current status. Corals throughout the Persian Gulf are in poor condition. Large areas were impacted by coral bleaching in 1996, 1998 and 2002. Recovery has occurred, but has been slow, particularly on reefs close to population centers. Measures of live coral cover in the Gulf are typically only 5 to 10 percent of the total reef surface. A bleaching

FIGURE 5.1. REEFS AT RISK IN THE MIDDLE EAST

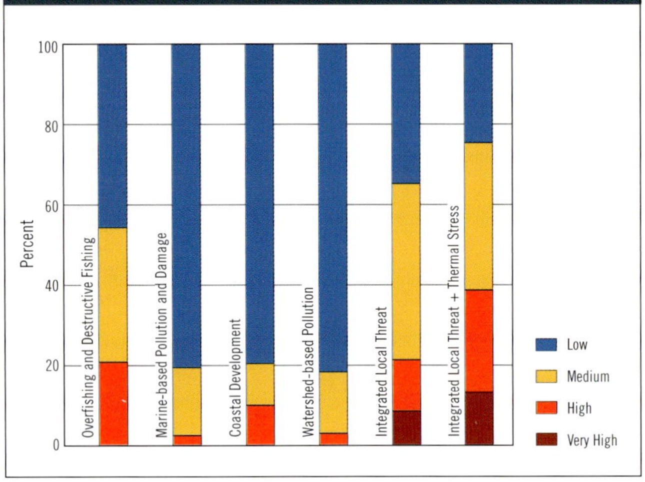

event in 2007 affected the reefs of Iran, but recovery has been good. In contrast to these stresses, the Red Sea and Gulf of Aden have good coral cover and have probably suffered less from coral bleaching than any other major reef region.

Overall results. Nearly two-thirds of the reefs in the region are at risk from local threats. The greatest pressure is in the Persian Gulf, where more than 85 percent of reefs are considered threatened, while the figure for the Red Sea is just over 60 percent. Areas of low threat in the central western Red Sea and along the northern Red Sea coast of Saudi Arabia may be some of the most extensive areas of reefs on the continental margin under low threat anywhere outside of Australia. The addition of past thermal stress increases the overall threat levels in the region to more than 75 percent and broadly captures the observed patterns of intense and destructive bleaching in the Persian Gulf, with relatively minor impacts in the Red Sea.

Overfishing dominates the local threat statistics, affecting 55 percent of reefs. Coastal development is spatially limited, but has grown considerably since 1998. Watershed-derived impacts are low compared with other regions, due in large part to the lack of runoff from the land. Marine-based pollution affects one fifth of reefs—a relatively high level for this threat, but still likely to be an underestimate. The threat analysis picks up the very heavy shipping traffic in the Red Sea, but shows little impact from the oil and gas industry on the coral reefs of the Persian Gulf. Although the major fields, pipelines, and shipping routes lie at some distance from most reefs, background levels of pollution are high throughout the Gulf, and probably affect much wider areas than our findings suggest.

Thermal stress and ocean acidification are projected to increase threat levels to nearly 90 percent by 2030, while by 2050 these climate change impacts, combined with current local impacts, will push all reefs to threatened status, with 65 percent at high, very high, or critical risk.

Conservation efforts. Only 12 percent of the region's reefs are within protected areas, many of which are in Egypt. About 50 percent of Egypt's reefs are inside MPAs and all of these MPAs are considered at least partially effective. These protected areas have likely played an important role in maintaining healthy reefs and reducing the impact of the burgeoning tourism industry over extensive areas of the Egyptian coast.

BOX 5.1 REEF STORY

Persian Gulf: The Cost of Coastal Development to Reefs

Coral reefs in the Persian Gulf have evolved to survive some of the highest temperatures and salinities on Earth. However, they are threatened by massive coastal and offshore development, which has caused a serious decline in associated habitats, species, and overall ecosystem function in the region. The key to stemming the decline from overdevelopment lies in greater regional-level coordination and a longer-term, holistic outlook for the gulf as an ecosystem. These approaches will help to ensure both the ecological and economic sustainability of the gulf into the future. *See full story online at www.wri.org/reefs/stories.*

Story provided by David Medio of the Halcrow Group Ltd.

MAP 5.2. REEFS AT RISK IN THE INDIAN OCEAN

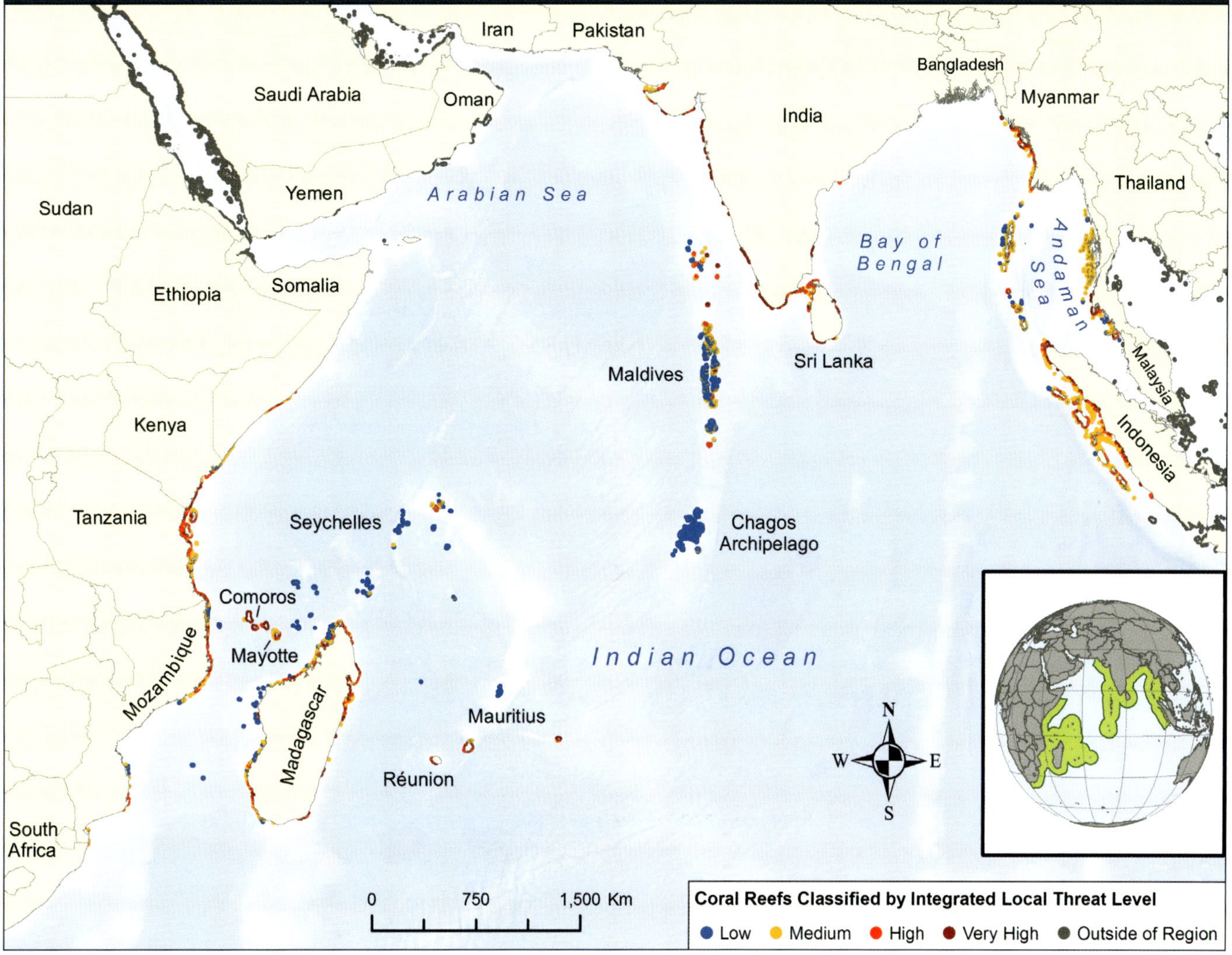

INDIAN OCEAN

The region. Stretching from East Africa to Sumatra, the Indian Ocean basin has extensive reefs (31,500 sq km) that are concentrated in three broad areas. The western Indian Ocean includes continental reefs and also the Seychelles, Comoros, and Mascarene oceanic islands. Deep oceans separate these from the vast reef tracts along the Chagos-Laccadives Ridge, including the Maldives. In the east, reefs encircle the Andaman Sea, including India's Andaman and Nicobar Islands, as well as the islands and complex mainland coasts of Myanmar and Thailand. Reefs are far less abundant around the Indian subcontinent, although there are important areas in the Gulf of Mannar and southern Sri Lanka.

Biodiversity. This region has 13 percent of the world's coral reefs. The eastern reefs are closely associated with the highly diverse reefs of Southeast Asia. Further west, the reefs are more isolated and boast many unique species, including between 30 and 40 percent of parrotfish and butterflyfish species.[158, 159]

People and reefs. Although many reefs in this region are remote from large human populations, more than 65 million people live on the coast within 30 km of a coral reef,[160] and many are highly dependent on these ecosystems for food, income, and coastal protection. In the Maldives, in particular, the islands themselves are built from reefs and the people depend heavily on fishing and on tourism. The same economic pillars of fishing and tourism are evident across the region. In countries such as the Seychelles, the Maldives,

FIGURE 5.2. REEFS AT RISK IN THE INDIAN OCEAN

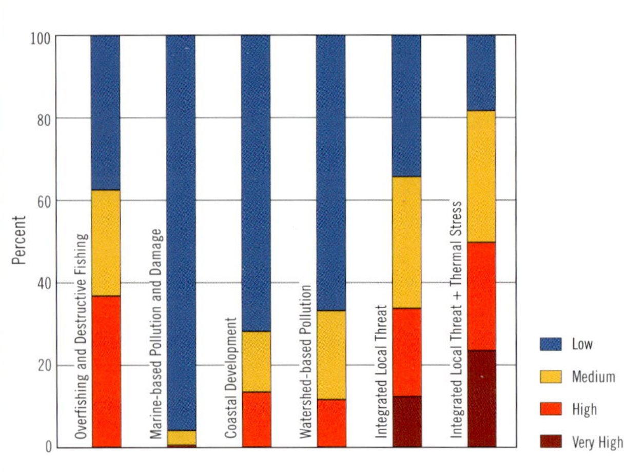

Thailand, Kenya, and Tanzania, reef-based tourism makes a critical contribution to the economy. These and other countries still rely heavily on fishing the reefs for subsistence and for income from sales to local markets.

Current status. The devastating bleaching of 1998 hit this region harder than any other.[161] In the Maldives, Chagos, and Seychelles, more than 80 percent of corals suffered complete mortality.[90, 162, 163] New studies of this event suggest that bleaching mortality was not simply correlated with temperature, but was influenced by patterns of historic variability in temperature.[149, 161] This may be of considerable importance in understanding and predicting future bleaching impacts. Further bleaching was recorded in 2001 and 2005 in these and other areas. Despite this, the region has also provided many examples of rapid recovery,[98, 99, 102] although such apparent recovery may hide underlying ecological changes, with some species not recovering as quickly or at all.[99, 164] In mid-2010, another coral bleaching event was reported from the Andaman Sea, with mortalities reaching 80 percent in some species. [165]

The Indian Ocean tsunami of 2004 affected large areas, including northern Sumatra, Thailand, the Andaman and Nicobar Islands, and Sri Lanka. Damage to reefs was localized, but coral at Car Nicobar (the northernmost of the Nicobar Islands) suffered more than 90 percent mortality. Here and in other areas, reefs are now recovering quickly, but in a few places around the Andaman and Nicobar Islands and northwest Sumatra, tectonic shifts in 2004 and 2005 caused significant sinking of coastal land in some places and uplift of reef above the water elsewhere, with the latter in particular killing wide areas of coral. [166]

Overall results. More than 65 percent of reefs in the Indian Ocean are at risk from local threats, with one-third rated at high or very high risk. Closer examination reveals a sharp focus of threatened areas along continental shores where more than 90 percent of reefs are threatened.

The single biggest threat is overfishing, which affects at least 60 percent of coral reefs, especially on the densely populated coastlines of southern India, Sri Lanka, southern Kenya, Tanzania, Thailand, and Sumatra. Dynamite fishing in this region is a localized problem, occurring mainly in Tanzania (Box 3.5). Watershed-based pollution is also a problem, especially in Madagascar, where extensive deforestation has led to massive erosion and siltation in many coastal areas.

Around the oceanic islands, the situation appears to be better. However, many of the reefs around the Andaman and Nicobar Islands, which were considered low risk in 1998, are now threatened. This is largely driven by growing populations, immigration, and tourism, as well as the impact of sediments following forest clearing. Similarly, the Maldives have shown a notable increase in threat since 1998, largely linked to overfishing. This may reflect the large population growth in this country, with a 10 percent increase in population between 2000 and 2006.[167] Although Maldivian fisheries largely target deep-water species such as tuna, they still depend on bait fish caught on the reefs, and there may be growing pressure both for home markets and the high-value export market for groupers. [168] Even among the low-risk, remote island reefs, some pressures are not captured in the model, notably the targeting of high-value species for live reef food fish trade with Asia in the Maldives and Seychelles, and the illegal capture of sharks and sea cucumbers from Chagos and elsewhere. [57, 169]

The integration of past thermal stress pushes the threat level on reefs beyond 80 percent, but even this may be an underestimate given the profound impact of the 1998 mass bleaching event on corals in most of the region. In a few places, patterns of recovery appear to be inversely correlated with local stress, with better recovery in areas where other

BOX 5.2 REEF STORY

Chagos Archipelago: A Case Study in Rapid Reef Recovery

The vast reef systems of the Chagos Archipelago are the most geographically isolated in the Indian Ocean and are far from most human influence, other than a large military base in the south. Chagos lost about 80 percent of its shallow and soft corals following severe bleaching in 1998.[90] Since then, and despite further bleaching in 2003 and 2005, there has been a remarkable recovery, highlighting the potential resilience of reefs to climate change where other human stresses are reduced or absent.[99] *See full story online at www.wri.org/reefs/stories.*

Story provided by Charles Sheppard of the University of Warwick.

stressors are more limited, such as the Chagos Archipelago[170] and the Maldives.[98] Slower or more inconsistent recovery has occurred in places such as the northern Seychelles, where reefs are subject to continuing ecological stress driven by other ongoing human impacts.[171] Such patterns are not ubiquitous, however, and do not appear to hold true at finer resolutions: one regional study was unable to find any clear correlation of improved recovery inside versus outside strict no-take marine protected areas.[172]

By 2030, projections suggest that climate-related threats will increase overall threat levels to more than 85 percent. Particularly dramatic changes are predicted off Madagascar and Mozambique, where threats of acidification and thermal stress coincide, although it is possible that the degree of resistance offered by past thermal history in these areas may ameliorate such patterns slightly.[161] Even the Chagos Archipelago, which currently has a low local threat, is projected to be threatened by 2030. By 2050 all areas will be considered threatened from the combination of local and climate-related threats, and most will fall under high risk from thermal stress and moderate risk from acidification. By this time, roughly 65 percent of reefs are projected to be at high, very high, or critical threat levels.

Conservation efforts. About 330 MPAs are established in this region, covering 19 percent of the coral reefs. Effectiveness assessments obtained for 58 percent of these MPAs concluded that one-quarter were considered ineffective, with just under half being partially effective (see Chapter 7.) A few of these sites in the more heavily populated parts of Kenya, Tanzania, and the Seychelles reduce the threat of overfishing in our model, and are helping to maintain healthy reefs in these places.[173, 174] In April 2010, the government of the United Kingdom declared an MPA to cover most of the Chagos Archipelago, and commercial fishing was formally ended on November 1, 2010. Although the site is patrolled and has relatively low pressures, we only marked it as partially effective, largely because of its unclear present and future legal status.[175] Chagos is presently the largest MPA in the world, adding almost 2,600 sq km of reef to the total MPA coverage. The Maldives still have very low levels of MPA coverage; however, the state government and fishing industry are making considerable progress in developing sustainable management of their offshore fisheries. In 2009, a national ban on nearshore shark fishing was introduced, a decision recognizing that the harvest was not sustainable, and was influenced by the considerable value placed on shark sightings by visiting tourists.[176]

SOUTHEAST ASIA

The region. Southeast Asia has the most extensive and diverse coral reefs in the world. They make up 28 percent of the global total (almost 70,000 sq km), concentrated around insular Southeast Asia, where fringing reefs predominate, and supplemented by barrier reefs such as the extensive Palawan Barrier Reef in the Philippines. Small but significant oceanic atoll and platform formations are also present, notably in the South China Sea. Most of the eastern half of this region lies in deep oceanic waters, which are of consid-

MAP 5.3. REEFS AT RISK IN SOUTHEAST ASIA

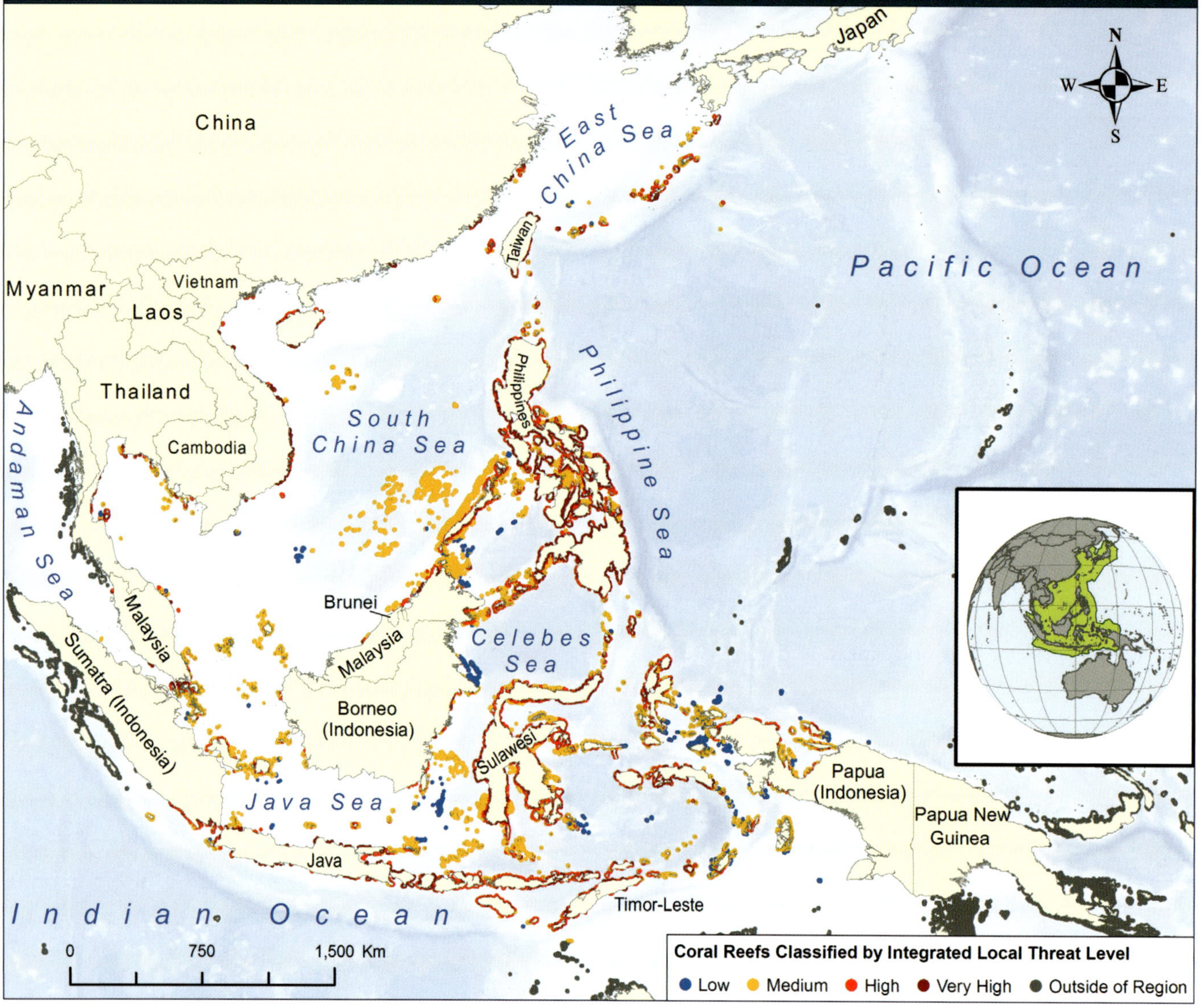

erable importance for reef health by stabilizing temperatures, diluting pollutants, and removing sediments.

Biodiversity. The reefs from the Philippines and east coast of Borneo across to Papua make up the western half of the Coral Triangle, the region with the highest diversity of corals, fish, and other reef species anywhere in the world. [177, 178] Diversity decreases in the shallow and sediment-rich coastlines of the Java Sea, and decreases further still in higher latitudes as waters become cooler. Even so, reefs thrive into southern Japan (the most northerly reefs after Bermuda) thanks to the warming influence of the Kuroshio Current. [179] This region is also host to some of the most extensive and diverse areas of mangroves and seagrasses anywhere in the world. [180, 181] The ecological connections between these ecosystems and coral reefs are important—both mangroves and seagrasses are known to support very high densities of a number of juvenile reef fish and likely offer enhanced survival for these individuals compared to those living in other habitats.[182]

People and reefs. Human populations are dense across much of the west of this region, including the Philippines and western Indonesia. More than 138 million people in Southeast Asia live on the coast within 30 km of a coral reef,[183] which is more than in all of the other coral reef regions combined. Fish, including reef fish, form a major part of the diet even in urban populations; across the region,

FIGURE 5.3. REEFS AT RISK IN SOUTHEAST ASIA

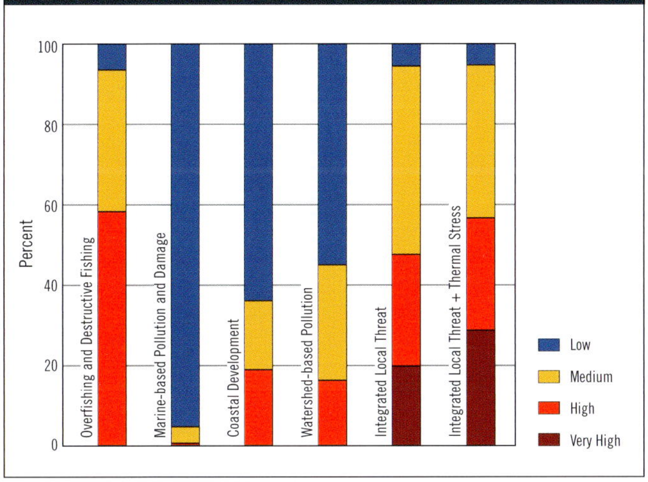

fish and seafood provide an average of 36 percent of dietary animal protein.[184] Deforestation has transformed wide areas, greatly adding to erosion and sedimentation problems in coastal waters. Mangrove losses have also been greater here than anywhere else in the world, linked to the massive expansion of aquaculture, especially in western parts of the region.[185] The loss of these natural filters has exacerbated sediment and pollution impacts on coastal coral reefs.

Current status. Overfishing has affected almost every reef in Southeast Asia, including areas remote from human habitation. Destructive fishing (blast and poison fishing) is rampant in this region, and although illegal, represents a major enforcement challenge. Sedimentation and pollution from land are significant, while coastal development is a growing threat. This region has suffered less than others from past coral bleaching events,[47] although medium-to-severe bleaching was recorded at a number of locations in mid-2010.[186]

Overall results. The reefs in this region are the most threatened in the world. About 95 percent are at risk from local threats, with almost half in the high and very high threat categories. The few places that are in the low-threat category are located in the more sparsely populated eastern areas.

The greatest threat is unsustainable fishing, which affects virtually all reefs. Destructive fishing alone affects at least 60 percent of reefs in the region. Very high human populations in most areas are driving fishers toward remote reefs to supply large regional markets. Demand from East Asian markets has a marked additional influence on the overfishing trend, often encouraging illegal fishing for shark fins, sea cucumbers, and live reef food fish on even the most remote reefs.

Coastal development is variable, but dense populations around the mainland continental shores, the entire Philippine Archipelago, and around Java and Sulawesi in Indonesia affect almost all reefs in those areas. Watershed-based pollution is even more widespread, not only in densely populated areas, but also around the Lesser Sunda Islands and Papua, where deforestation and agricultural expansion are increasing soil erosion and sedimentation. Although mangroves still play an important role in reducing

BOX 5.3 REEF STORY

Indonesia: People Protect Livelihoods and Reefs in Wakatobi National Park

Many larger reef fish such as groupers and snappers travel long distances to spawn in dense aggregations. Fishers often target such gatherings, rapidly decimating the population and simultaneously reducing the natural restocking of the reefs with new fish larvae. Preventing fishing on these spawning aggregations is a considerable challenge, but in Wakatobi National Park a growing awareness of declining fish populations has helped to fuel community-led initiatives, in collaboration with park authorities, to close fishing on what some locals have termed "fish banks." These measures have reversed the decline in the number of spawning groupers and snappers, with the expectation that recovery of entire populations will follow. *See full story online at www.wri.org/reefs/stories.*

Story provided by Joanne Wilson and Purwanto of The Nature Conservancy, Indonesia; Wahyu Rudianto of Wakatobi National Park Authority; Veda Santiadji of the World Wildlife Fund, Indonesia; and Saharuddin Usmi of KOMUNTO, Wakatobi National Park.

watershed pollution in many areas, loss and degradation of mangroves, notably from conversion to aquaculture in the Philippines and western Indonesia, have greatly reduced this important function.[180]

At such high levels of local threat it seems remarkable that many reefs still have good coral cover and high fish diversity, especially when compared to other areas of extensive high threat, such as the Caribbean. A number of factors help to promote reef survival in this region. First, major coral bleaching-related mortality had not affected large areas, at least until 2010. Second, there have been considerably fewer impacts from diseases than in other regions. Third, major ocean currents, notably the Indonesian Through-Flow, which passes through the eastern islands of the region, may be removing pollutants and supporting connectivity between reefs, with rapid and continuous movements of larvae from place to place, enhancing resilience and recovery from such localized impacts as blast fishing. Finally, it is also possible that the region's high levels of diversity may increase resistance or resilience of coral reefs.[187]

The addition of past thermal stress does not alter the proportion of reefs rated as threatened, although it does increase the number of reefs rated at very high threat from about 20 to 30 percent. This relatively minor influence may be linked to a relatively low incidence of thermal stress over the past decade compared to other regions, a situation which appears to be changing. The future threats from both warming and acidification will compound the problems in this region: we project that by 2030, 99 percent of reefs will be threatened, with the vast majority (more than 80 percent) at high, very high, or critical levels. In 2050, all reefs will be threatened, with about 95 percent at the highest levels.

Conservation efforts. Nearly 600 protected areas cover 17 percent of the region's reefs. Unfortunately, of the 339 that were rated, 69 percent were classified as not effective and only 2 percent as fully effective, covering a mere 16 sq km of coral reef. Nonetheless, there have been some important developments in the region. Apo Island in the Philippines stands testimony to the considerable value of MPAs to local communities, who have benefited for almost 30 years from increased fish catches due to the presence of a strict no-take zone.[188] Komodo, in Indonesia, has also benefited considerably from international support; and although there are still challenges, blast fishing and other pressures have been greatly reduced. Perhaps the most important trend has been the growth of locally managed marine areas, notably in the Philippines (see Chapter 7).[189] These marine areas represent a highly dispersed network of refuges that may be critical for reef survival and recovery in future years.

AUSTRALIA

The region. Joining the Indian and Pacific Oceans, and with extensive northern coastlines adjacent to Southeast Asia, Australia is home to more coral reefs than any other single nation—42,000 sq km, or 17 percent of the global total. Numerically, most of Australia's reefs form part of the vast Great Barrier Reef, which stretches over 2,300 km in length, and alone covers nearly 37,000 sq km of coral reef area.

The northern coasts feature scattered patch reefs and fringing reefs around offshore islands, becoming more widespread further west and including atolls and banks on and beyond the continental shelf margin. Coastal reefs are less common, but along the North West Cape, Ningaloo is one of the world's largest continuous fringing reefs, at 230 km long. There are also several oceanic reefs, notably around the Christmas and Cocos (Keeling) islands in the Indian Ocean, the reefs of the Coral Sea in the Pacific, and also some of the southernmost reefs in the world on Lord Howe Island.

Biodiversity. Spanning two oceans, Australia's reefs embrace considerable diversity, with characteristics of Indian Ocean species in the west and Pacific species in the east. Spanning significant latitudes means that these reefs offer excellent examples of the natural gradients in the diversity of corals and other species, with diversity decreasing as one moves toward higher latitudes, away from the tropics.

People and reefs. Most of Australia's reefs lie far from large human populations. Even where there are population centers, notably along parts of the coast of Queensland, the reefs generally lie more than 30 km offshore. The exception is in the Cairns region, where fringing reefs line much of the coast and platform reefs lie as little as 20 km offshore. Australia has the lowest coastal population densities of any region in this study, with only about 3.5 million people liv-

MAP 5.4. REEFS AT RISK IN AUSTRALIA

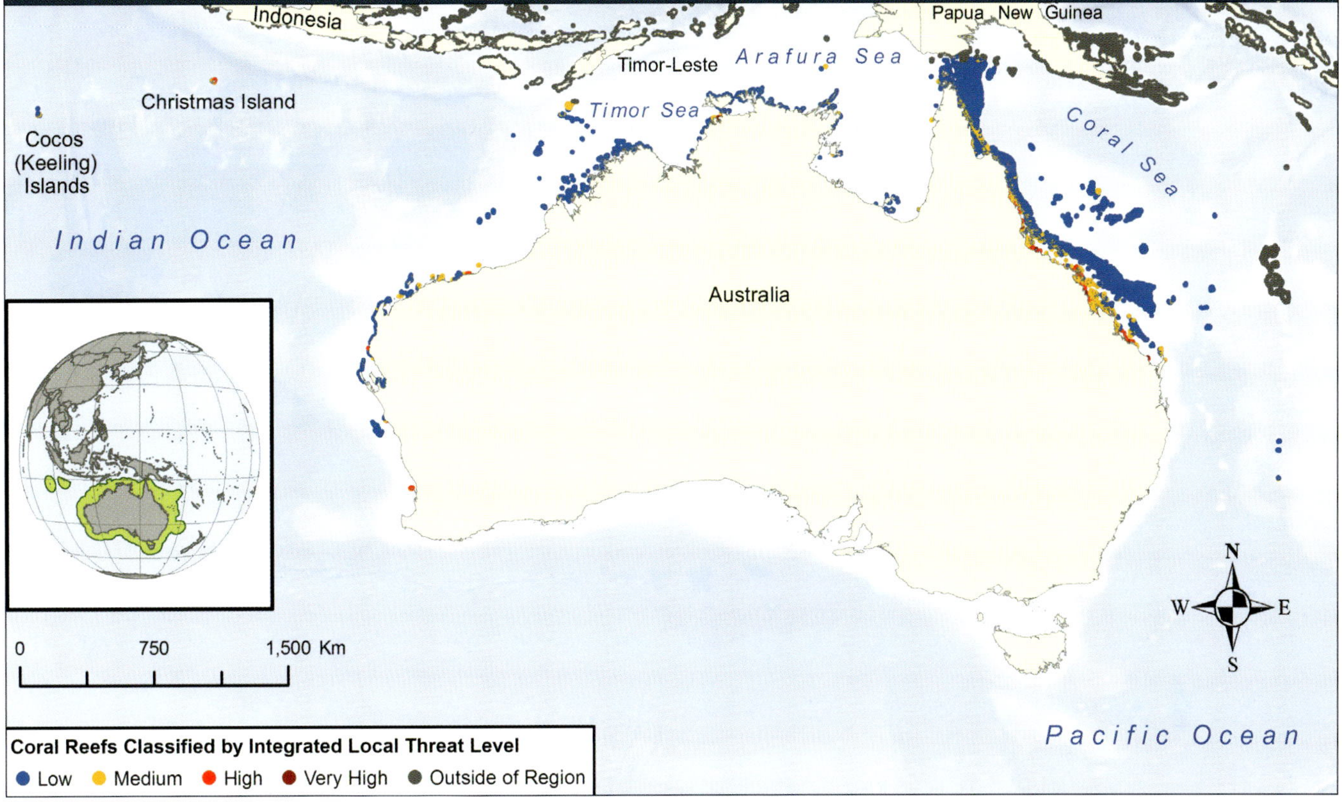

ing on the coast within 30 km of a coral reef.[190] Despite this, the reefs are an important resource. Tourism on the Great Barrier Reef is a critical part of the region's economy, generating US$5.2 billion in 2006.[191] Recreational fishing is a popular activity for locals and visitors alike, while important commercial fisheries target fish, sharks, lobsters, crabs, and prawns using a range of fishing gear, including lines, nets, pots, and trawls.

Current status. Detailed studies of Australia's reefs date back decades. Coral cover has been shown to fluctuate widely in part as a result of occasional devastating impacts from tropical cyclones, as well as outbreaks of crown-of-thorns starfish (COTS) and of the coral-eating snail *Drupella* (mainly in western Australia). Mass bleaching events, notably in 1998 and 2002, also damaged reefs. Despite these impacts, broad-scale, long-term trends appear to show stable coral cover, with upward trends in some areas.[73] The re-zoning of the Great Barrier Reef Marine Park in 2004 led to a significant expansion of strictly protected areas from 4 percent to 33 percent of the entire marine park. This expansion already appears to be having a significant positive influence, both on fish communities and on overall ecological resilience of biodiversity.[75, 192] A similar re-zoning in Ningaloo Marine Park in 2006 may offer valuable support for increasing or at least maintaining reef health.

Overall results. The reefs of Australia are the least affected by local threats of any region. About 15 percent are threatened by local stressors, with only about 1 percent at

FIGURE 5.4. REEFS AT RISK IN AUSTRALIA

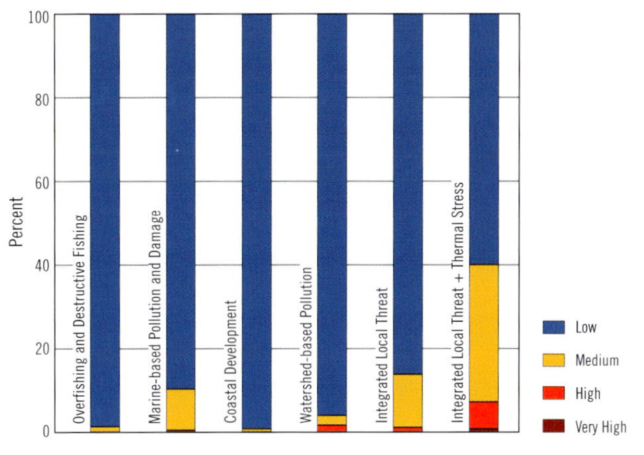

BOX 5.4. REEF STORY

Australia: Remaining Risks to the Great Barrier Reef

Australia's Great Barrier Reef is the world's largest coral reef ecosystem and is almost completely contained within a marine protected area. Despite recognition that it is one of the world's best-managed reefs, its long-term outlook is poor due to the anticipated impacts of climate change (that is, warming and acidifying seas). As in other areas, climate-related threats can be compounded by local threats originating outside the park, including coastal development, mining, and agricultural runoff, which cause poor-quality water to drain into the marine park. In response, national and state governments have developed a coastal water quality protection plan, and the Great Barrier Reef Marine Park Authority has launched the Reef Guardian program to collaborate with local governments, schools, and communities to use best practices within the watershed and to build resilience into the reef ecosystem in the face of climate change. *See full story online at www.wri.org/reefs/stories.*

Story provided by Jason Vains and John Baldwin of the Great Barrier Reef Marine Park Authority.

PHOTO: COMMONWEALTH OF AUSTRALIA (GBRMPA)

high or very high threat. This threat percentage is much lower than in the 1998 analysis (Figure 4.9), largely due to a more precise and conservative threat analysis method.

The threat analysis identified marine-based pollution and damage and watershed-based pollution as the major local threats to Australia's reefs. Marine-based pollution and damage is a moderate threat for 10 percent of Australia's reefs, largely driven by the relatively busy shipping lanes that traverse the Great Barrier Reef and bring many boats into relatively close proximity to coral reefs, particularly in more northern areas. In reality, shipping is strictly managed and the number of incidents is low; 54 major incidents have been recorded from 1985 to 2008, but the actual spatial footprints of these, from physical impacts to pollution, is still very small.[73, 193]

Watershed-based pollution from adjacent agriculture and forest clearance has been widely recorded in the Great Barrier Reef.[194] Our analysis suggests that watershed-based pollution threatens about 4 percent of Australia's reefs, including 2 percent at high threat. Although a small proportion, this includes virtually all of the nearshore reefs in the southern and central sectors of the Great Barrier Reef. These nearshore ecosystems not only host unique biodiversity, but are of particular importance to people. It is possible that this threat may in fact extend more broadly, since detailed studies of the Great Barrier Reef have shown slightly wider areas of impact.[73]

Overfishing and coastal development are each estimated to threaten only 1 percent of reefs, largely because of the great distances of most of the reefs from people. In reality, some impacts of fishing have been recorded even on remote reefs, and it seems likely that both recreational and commercial fishers travel further than the analysis suggests.[195]

Thermal stress on Australia's reefs has had a dramatic impact during the last ten years. When this is incorporated into the model, more than 40 percent of reefs are rated as threatened. Furthermore, the projections of future impacts from both warming and acidification suggest even more dramatic changes. Combined local and climate change impacts raise overall threat levels to nearly 90 percent by 2030, with 40 percent of reefs rated at high, very high, or critical threat. Some of the most highly threatened reefs are in the northern Great Barrier Reef, but by 2050 more than 95 percent of Australian reefs are rated as threatened, including most of the Great Barrier Reef. Although the outlook is bleak, this region has fewer reefs in the *very* high threat category (less than15 percent) than any other region.

Conservation. About three-quarters of Australia's coral reefs fall within marine protected areas. This includes 30,000 sq km (12 percent of the world's coral reefs) in the Great Barrier Reef Marine Park. There is a high level of active management within many of these sites, including specific plans to control tourism and regulations governing commercial and recreational fishing. Recent re-zoning of the Great Barrier Reef and Ningaloo marine parks has classified one-third of each site as strict, no-take zones. Such changes were made only after long periods of consultation with all relevant stakeholders. Efforts also extend beyond MPA boundaries and there are active policy and management processes to reduce watershed-based pollution on the Great Barrier Reef, notably in reducing runoff of sediments, nutrients, and pesticides from the land.

PACIFIC

The region. The Pacific Ocean spans almost half the globe, from Palau in the west to the coastline of Central America in the east, and holds more than a quarter of the world's coral reefs, nearly 66,000 sq km. Most of these reefs are found among the three major island groups of the western Pacific. In the northwest, Micronesia consists of several archipelagos dominated by coral atolls, but with a few high islands of volcanic origin. The Melanesia group in the southwest has the largest land areas: stretching from Papua New Guinea in the west to Fiji in the east, most reefs are fringing and barrier formations, including New Caledonia's 1,300 km barrier reef, second only to Australia's Great Barrier Reef in length. The Polynesian islands occupy a vast area of the central Pacific, including Tonga, French Polynesia, and Hawaii to the north. Here, coral atolls predominate, with a few high volcanic islands.

In the eastern Pacific, coral reefs are rare, with one small raised atoll (Clipperton) and the remainder being small patch, bank, and fringing reefs. Reef development in most of the region is inhibited by a combination of factors: cold water upwellings, high variability of temperatures between years, and large volumes of freshwater runoff and sediment in many areas.

Biodiversity. Papua New Guinea and the Solomon Islands make up the eastern half of the Coral Triangle, the global center of reef diversity with more species of fish, corals, and other groups than anywhere else. [178, 196] Moving away from this region, biodiversity decreases gradually both to the east and toward higher latitudes. The easternmost islands, including the Pitcairn Islands, the Marquesas Islands (in French Polynesia), and Hawaii all have relatively low diversity, but their isolation has supported the evolution of large numbers of endemic species.[158] The eastern Pacific reefs have very low diversity, but most of the species found there are unique. [197]

People and reefs. More than any other region, the people of the western Pacific are closely connected to coral reefs. At least 7.5 million people in the Pacific islands live in coastal areas within 30 km of a coral reef, representing about 50 percent of the total population.[198] For many, reefs are a critical mainstay in supporting local fisheries, while in some areas reefs are also supporting export fisheries and tourism. The importance of reefs is heightened by a lack of alternative livelihoods, particularly in the many very small island nations. Sea level rise poses a considerable threat in many of these same countries, where most or all of the land consists of coral islands. Rising seas can infiltrate groundwater—killing crops and threatening freshwater supplies—while inundation during the highest tides and severe storms is already a threat in some areas.[199] Healthy reefs offer some resistence to coastal erosion, and supply the sand and coral rock needed to build and maintain coral islands. However, corals may not be able to continue to provide these services under accelerating levels of sea level rise, especially in combination with degradation of the reefs themselves.[132, 200]

The connection between people and coral reefs in the eastern Pacific is more limited. Although fishing is widespread, it more typically targets mangrove-associated species, and pelagic species where the continental shelf is narrow.

Current Status. Although large areas of the Pacific still have relatively healthy reefs with high coral cover, this is changing. Overfishing is fairly widespread, including the targeted capture of sharks and sea cucumbers for export to East Asian markets, as well as chronic overfishing in some places.[201] This has been exacerbated in countries such as the Solomon Islands by the breakdown of traditional management approaches. In Papua New Guinea, sedimentation

MAP 5.5a. REEFS AT RISK IN THE WESTERN PACIFIC

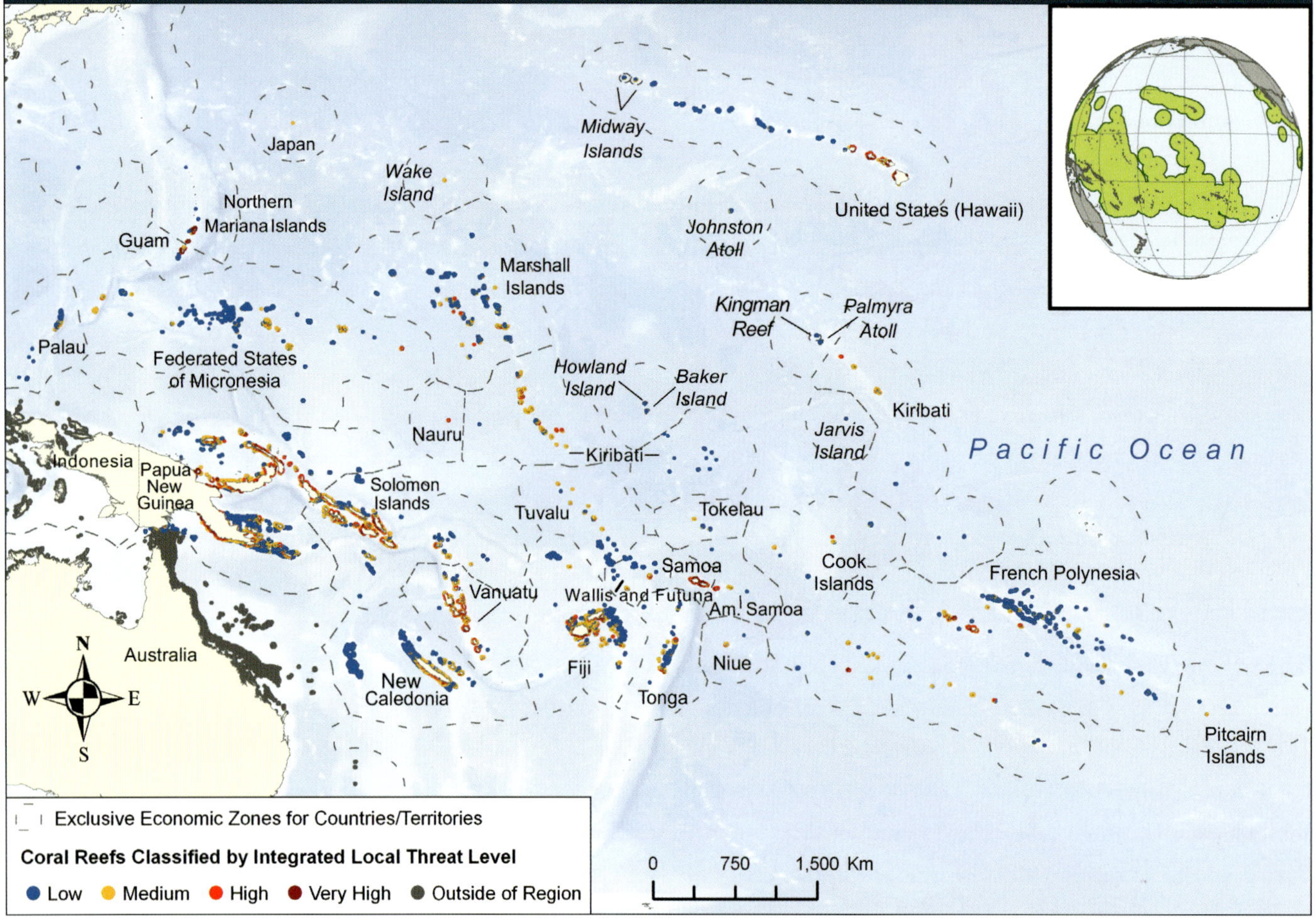

and pollution from inland areas are a threat to reefs. Natural impacts have also affected some areas, including outbreaks of COTS as far afield as Papua New Guinea, Pohnpei, the Cook Islands, and French Polynesia. Seismic activity, including a tsunami and uplift of reefs and islands by up to 3 meters, caused considerable damage in the Solomons in 2007. On a more positive note, many countries have established very large numbers of locally managed marine areas (see Chapter 7), with local ownership and management. Coral bleaching impacts in Micronesia included severe bleaching in Palau (1998) and Kiribati (2002–03 and 2005). Recovery has been good in at least some of these areas, although with some changes to the dominant corals noted in Tarawa, Kiribati.[202] Significant bleaching was also recorded in Palau in late 2010, with records of elevated sea surface temperatures through large parts of Micronesia.

MAP 5.5b. REEFS AT RISK IN THE EASTERN PACIFIC

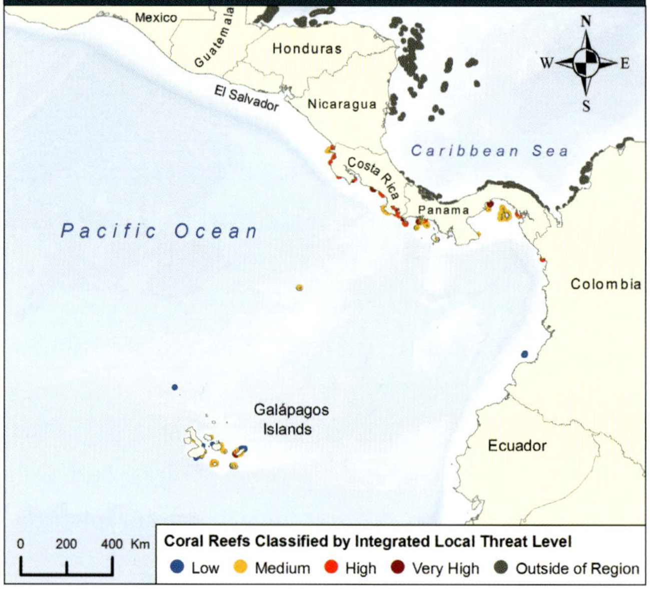

BOX 5.5 REEF STORY

Line Islands: A Gradient of Human Impact on Reefs

The Line Islands, a chain of a dozen atolls and coral islands in the central Pacific Ocean, are home to some of the most remote and pristine coral reefs on Earth. The uninhabited atolls of Millennium and Kingman provide a glimpse of coral reefs before human impacts, including incredible coral formations and an abundance of predators. The populated atolls of Tabuaeran and Kiritimati and the island of Teraina, on the other hand, show a decline in reef health due to overfishing and pollution. Recent studies of the atolls reveal a strong association between increasing human population and ecosystem decline. They show how human influence is the most paramount determinant of reef health, and add valuable evidence to the growing understanding of how minimizing human impacts on reefs may increase their resilience to global climate change. *See full story online at www.wri.org/reefs/stories.*

Story provided by Enric Sala of the National Geographic Society.

In the eastern Pacific, threats are highly variable. These reefs suffered the earliest known mass bleaching and mortality of any region, with widespread losses linked to an El Niño event in 1982–3. Sedimentation is a problem on some coastal reefs, notably in Costa Rica. Corals have also suffered from algal overgrowth by an invasive seaweed (*Caulerpa sertularoides*) and from red tides—blooms of toxic algae living in the plankton, which may be short-lived but can lead to high levels of toxic compounds in the food chain, threatening fish and even human consumers. Such algal growth may be exacerbated by high nutrient levels in coastal waters, linked to agricultural and urban pollution.

FIGURE 5.5. REEFS AT RISK IN THE PACIFIC

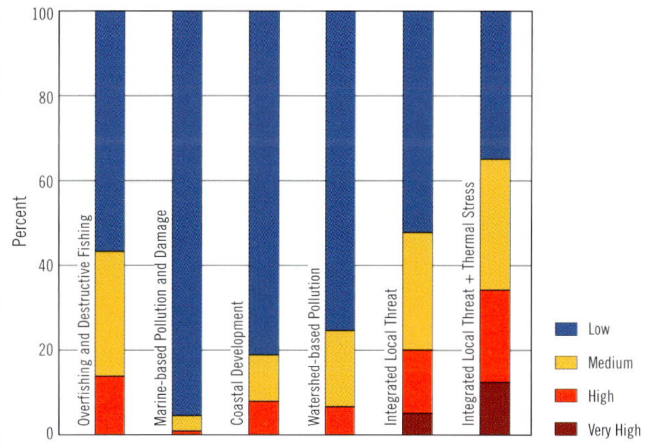

Overall results. After Australia, the Pacific is the least threatened region, with slightly less than 50 percent of reefs classed as threatened, and only 20 percent at high or very high threat. Most of the threatened reefs in the region are associated with large islands and areas of higher population, concentrated in Melanesia, but also in Hawaii, Samoa, and the Society Islands (in French Polynesia). The inclusion of past thermal stress raises the percentage of threatened reefs to more than 65 percent.

Overfishing is the largest threat, linked to densely settled areas not only around the larger islands, but also in some smaller archipelagos, including parts of Micronesia. Watershed-based pollution is limited to high islands, but is nonetheless widespread and affects a quarter of all reefs. In many areas this is linked to forest clearance and erosion, but open-cut mining is also a significant source of sediments and pollutants, notably in Papua New Guinea (copper and gold) and New Caledonia (nickel). Coastal development affects almost a fifth of reefs, notably in Hawaii, Fiji, Samoa, and the Society Islands.

In the eastern Pacific, the analysis suggests significant threat from watershed-based pollution in many areas, notably the continental coasts of Costa Rica. Overfishing is widespread in many of the same areas as well as around some of the offshore islands, including the Galapagos.

Future climate change impacts are projected to bring the proportion of threatened reefs up to 90 percent by 2030. Around Papua New Guinea, the Solomon Islands,

BOX 5.6 REEF STORY

New Caledonia: Reef Transplantation Mitigates Habitat Loss in Prony Bay

Prony Bay in southern New Caledonia, located 1,200 km east of Australia, is renowned for its exceptional reef communities. In 2005, a nickel mining corporation, Vale Inco NC, agreed to fund the transplantation of corals to compensate for reef habitat lost during the construction of a port. After three years, 80 percent of the individual coral transplants were still alive and in good health. Fish have also colonized the restored site and the surrounding reef appears to have a more diverse and denser reef community. With adequate resources and favorable conditions, such transplantation may offer a successful last resort to save an otherwise certain net loss of reef habitat. *See full story online at www.wri.org/reefs/stories.*

Story provided by Sandrine Job, Independent Consultant (CRISP Programme).

PHOTO: SANDRINE JOB

and Vanuatu, the combined impacts of acidification with thermal stress are projected to push many reefs into the very high or critical threat categories. By 2050, almost all reefs in the Pacific are rated as threatened, with more than half rated at high, very high, or critical levels. Parts of the South Pacific, such as southern French Polynesia, are rated at slightly lower risk.

Conservation. We identified more than 920 MPAs across the Pacific, covering about 13 percent of the region's reefs. Perhaps the most important and distinctive regional trend has been the recent rapid growth of local protection, notably through the establishment of locally managed marine areas. Such sites are established by local communities, with the support of partners such as NGOs or governments, and include areas of permanent or temporary closure, as well as more specific restrictions on fishing methods, target species, or access to the fishery. Our map includes 650 such sites, with the largest numbers in Fiji, Papua New Guinea, Samoa, and the Solomon Islands, but it is likely that many other sites have gone unrecorded. The influence of these sites on reef health and conservation is variable. Many are very small, and they have varying levels of effectiveness. However, their proximity to areas of overfishing pressure may offer a highly effective tool to build resilience and act as refuges. At the other extreme, the Pacific also contains most of the world's largest marine protected areas, including the Phoenix Islands Protected Area in Kiribati, several sites around U.S. territories (Papahānaumokuākea, Rose Atoll, the Mariana Trench, and the Pacific Remote Islands Marine National Monuments), and the Galapagos Marine Park. Despite their size (combined, they are over 1.3 million sq km), these sites incorporate less than 5 percent of the region's reefs. Designated around remote places with few or no resident local populations, these MPAs provide only limited benefits to current reef health, but are an important safeguard against future threats, and may contribute to longer-term regional reef resilience.

ATLANTIC

The region. The Atlantic region includes 10 percent (26,000 sq km) of the world's coral reefs. These reefs are restricted to the western half of the Atlantic Ocean, mostly in the Caribbean Sea and the Bahamas Banks. Reef types include fringing and bank reefs, as well as a number of long barrier-like systems, notably around Cuba and off the coast of Belize. The Bahamas group, which includes the Turks and Caicos Islands, is a huge system of shallow banks with reefs on their outer margins. Far out in the Atlantic Ocean, Bermuda represents an isolated outpost and the most northerly coral reefs in the world, connected to the Caribbean by the warm Gulf Stream. An even larger gap separates the Caribbean reefs from a number of small reefs off the coast of Brazil.

Biodiversity. The diversity of reef species in the Atlantic is comparatively low. While there are more than 750 species of reef-building corals across the Indian and

Pacific Oceans, the Atlantic hosts less than 65.[203] However, the Atlantic species are unique—well over 90 percent of fish, corals, crustaceans, and other groups are found nowhere else. Brazil's reefs have even lower diversity, with strong Caribbean links but also quite a number of endemic species.[a]

People and reefs. The region is densely populated and politically complex, with many small-island nations across the Caribbean. In this region, about 43 million people live on the coast within 30 km of a coral reef.[204] With the political diversity comes considerable economic diversity, and while many countries are relatively wealthy, there is still direct dependence on reefs for food and employment in many areas. Tourism is a critical economic pillar for many nations, and for those with relatively poor agricultural or industrial sectors, it is one of the few available livelihoods. Most tourism is concentrated on the coast, a significant portion of which is directly reef-related, with snorkeling and scuba diving among the most popular activities in countries and territories such as the Bahamas, Cayman Islands, Turks and Caicos, Bonaire, and Belize. Even in locations where reef visitation is lower, reefs play a hidden role: providing food, protecting coastlines, and providing sand for beaches. This region is prone to regular and intense tropical storms, and numerous coastal settlements are physically protected by barriers of coral reefs, breaking the waves far offshore and reducing the effects of devastating flooding and erosion.

Current status. Corals across this region have been in decline for several decades,[140] and some studies have traced the declines to systematic overfishing going back centuries.[17, 195] Since the 1980s a major cause of reef decline has been the impact of diseases, notably affecting long-spined sea urchins (*Diadema antillarum*) and many corals. Urchins are

a. Although a few isolated corals and reef fish species are found on the isolated islands of the central Atlantic, and in a few places off the coast of West Africa, they do not build reef habitats and are therefore not considered in this analysis.

MAP 5.6a. REEFS AT RISK IN THE ATLANTIC/CARIBBEAN

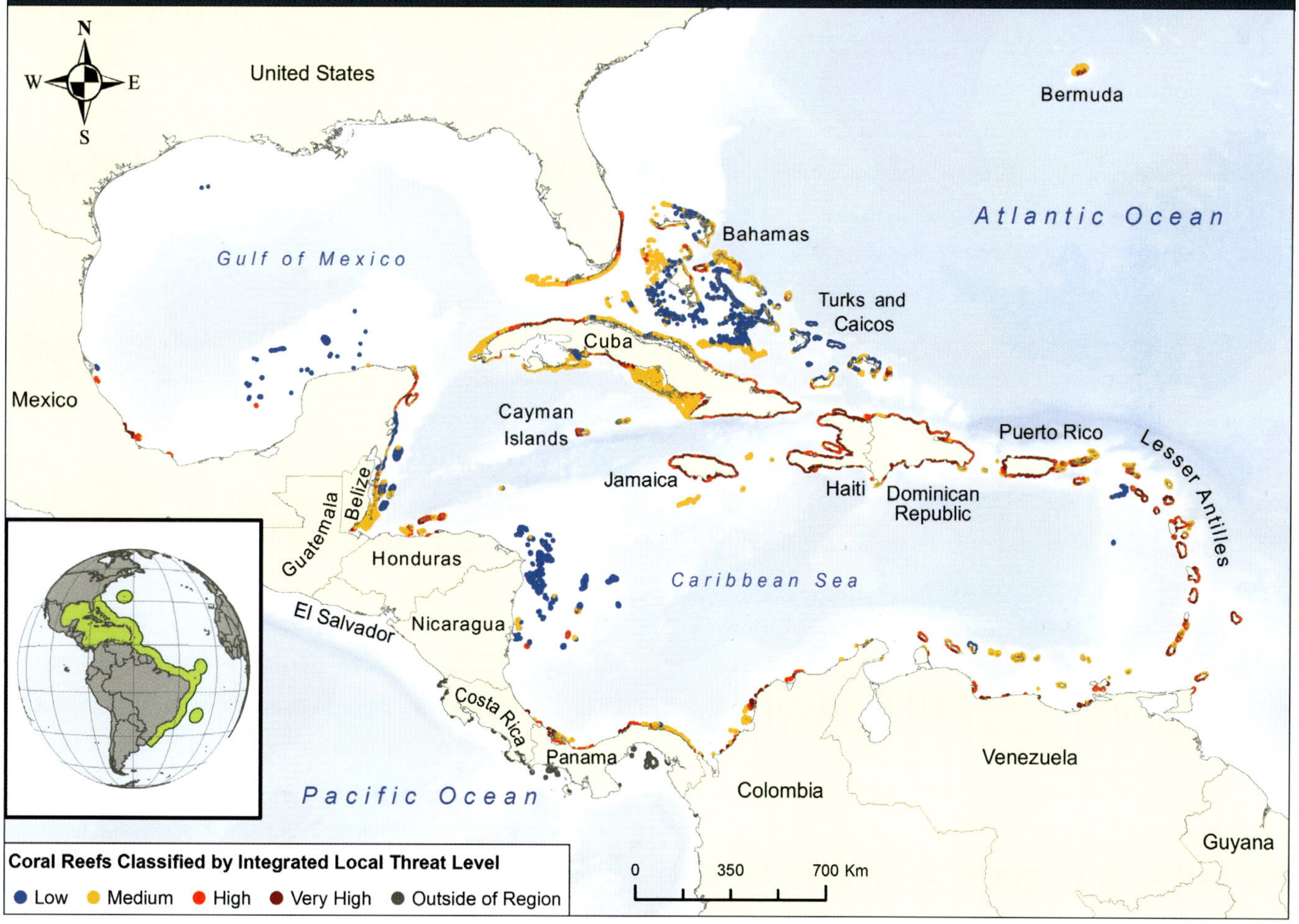

MAP 5.6b. REEFS AT RISK IN BRAZIL

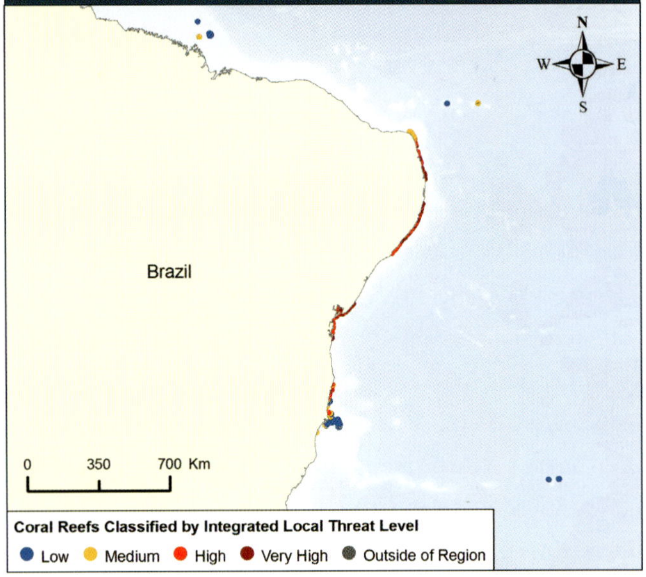

FIGURE 5.6. REEFS AT RISK IN THE ATLANTIC

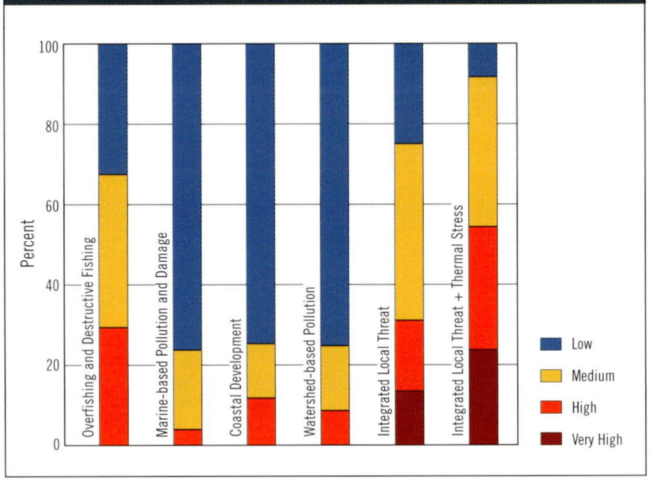

important grazers, feeding on algae and making space for corals. This role is particularly important in areas where overfishing has removed many grazing fishes. Coral diseases have led to the extensive loss of the two most important reef-building corals—staghorn and elkhorn. [205] Scientists are still debating the origins of these (and other) diseases, and whether human influence may be partly to blame. There is, however, strong evidence that coral diseases are more prevalent after coral bleaching events and that reefs subject to local human stressors such as pollution are also more vulnerable to disease.[22, 139, 206] Overfishing has affected almost all areas and Atlantic reefs have some of the lowest recorded fish biomass measures in the world. Overfishing occurs on virtually every reef: larger groupers and snappers are rare throughout the region, and on many reefs even herbivorous fish are much reduced.

The complex interplay of threats is better understood in the Caribbean than in many other places, partly as a result of intense, ongoing research in many countries. The co-occurrence of multiple threats is a particular problem. Reefs have survived heavy overfishing, but the combination of this threat with disease, hurricanes, pollution, and coral bleaching has been devastating for countries such as Jamaica, and for many areas in the Lesser Antilles. Coral bleaching caused considerable declines in coral cover across large parts of the region in 2005, but in many places the impact of the bleaching was exacerbated by physical damage from a series of large hurricanes; twelve hurricanes and eight tropical storms hit the northern Caribbean between 2004 and 2007, including some of the most powerful storms on record.[92] In some cases, as in the U.S. Virgin Islands in 2005, disease outbreaks following bleaching events have led to further large losses.[206] The overall loss of living coral cover in the region has led to a loss of reef structure. Since many species rely on the complex network of surfaces and holes that characterize living coral-rich reefs, such a decline has likely already impacted biodiversity and productivity across the region.[207]

Overall results. More than 75 percent of the reefs in the Caribbean are considered threatened, with more than 30 percent in the high and very high threat categories. Although high, we believe these figures may still be an underestimate given the many threats described above. Thus, while overfishing is rated as the most pervasive threat, affecting almost 70 percent of reefs, the reality may be even worse, with the only healthy reef fish populations being recorded from a small number of well-managed no-take MPAs. For example, in the Florida reef tract, coral cover has collapsed and pollution and overfishing remain widespread in all but a very small area of marine reserves.[104, 208]

Other pressures are also extensive, with marine-based pollution, coastal development, and watershed-based pollution each threatening at least 25 percent of reefs. Coral bleaching has also damaged many Caribbean reefs; the addition of past thermal stress in our analysis increases the overall threat to more than 90 percent of reefs, with almost 55 percent at high or very high threat. Evidence suggests that multiple impacts are driving greater declines, even more than the expected sum of the parts. Coral disease, although not built into the model, has perhaps been one of the greatest drivers of decline in this region; its influence exacerbated by the high levels of ecosystems impacts from other factors.

The maps show that the reefs considered to be under low threat are almost entirely in areas remote from large land areas, such as the Bahamas, the southern Gulf of Mexico, and the oceanic reefs of Honduras and Nicaragua. The insular Caribbean is particularly threatened: from Jamaica through to the Lesser Antilles, more than 90 percent of all reefs are threatened, with nearly 70 percent classified as high or very high threat. Most of these areas are affected by multiple threats, led by coastal development and watershed-based pollution. Brazil's reefs are similarly threatened, with only the offshore reefs remaining in low threat categories.[209]

By 2030, climate-related threats are projected to be high along the mainland coast from Mexico to Colombia, due to the confluence of thermal stress and acidification threats. Thermal stress is also projected to be high in the eastern Caribbean. Climate-related threats are projected to push the proportion of reefs at risk to 90 percent in 2030, and up to 100 percent by 2050, with about 85 percent at high, very high, or critical levels.

Conservation. The Atlantic region has 617 MPAs covering about 30 percent of the region's reefs. We were able to assess the effectiveness of about half of these; of that number, 40 percent (by reef area) were classified as ineffective, with only 6 percent by area (460 sq km) classified as fully effective. These very low effectiveness estimates reflect the immense challenges of establishing effective conservation when the pressures are so intense, management is costly, and full community engagement can be difficult to achieve. Despite these concerns, the existing MPAs may be helping: there is some evidence that even partial protection provides ecological benefits, including greater resilience to threats,[210] while the legal existence of sites can be a precursor to improving management.[211] Of course, even well-protected sites are still affected by regionwide problems in the Caribbean, and it is clear that major improvements in reef condition will require a broader array of management interventions.

BOX 5.7 REEF STORY

Florida: Marine Management Reduces Boat Groundings

Southeast Florida's extensive reefs lie close to three major sea ports and the damage from large boat groundings and dragging anchor cables can be considerable. Close to Port Everglades, where 17 such incidents were recorded in the 12 years to 2006, the problem has been ameliorated through a consultative process that led to the relocation of the port's main anchorage further from the reefs. *See full story online at www.wri.org/reefs/stories.*

Story provided by Jamie Monty and Chantal Collier of the Florida Department of Environmental Protection.

PHOTO: DAVID GILLIAM

Chapter 6. SOCIAL AND ECONOMIC IMPLICATIONS OF REEF LOSS

PHOTO: JOSHUA CINNER, ARC CENTER OF EXCELLENCE

Healthy coral reefs provide a rich and diverse array of ecosystem services to the people and economies of tropical coastal nations. Reefs supply many millions of people with food, income, and employment; they contribute significant export and tourism revenues to national economies; they perform important services such as protecting shorelines and contributing to the formation of beaches; and they hold significant cultural value for some coastal societies.[212] In many nations, reef ecosystem services are critically important to livelihoods, food security, and well-being. As a result, threats to reefs not only endanger ecosystems and species, but also directly threaten the communities and nations that depend upon them.

The relative social and economic importance of reefs is increased by the fact that many reef-dependent people live in poverty. Most of the world's inhabited coral reef countries and territories are developing nations.[213,214] Of these, 19 countries are also classified as least-developed countries (LDCs), due to a combination of low income, limited resources, and vulnerable economies.[215] Forty-nine reef nations are small-island developing states, where vulnerability is often compounded by high population densities, limited natural resources, geographic isolation, fragile economies, and susceptibility to environmental hazards such as hurricanes, tsunamis, and sea level rise.[216] For many reef nations, a shift toward more sustainable use of reef resources may offer valuable opportunities for poverty reduction and economic development.

This chapter examines the potential social and economic vulnerability of coral reef nations to the degradation and loss of reefs. In this assessment, we build on the findings of the threat analysis by examining where identified threats to reefs may have the most serious social and economic consequences for reef nations. We represent vulnerability as the combination of three components: exposure to reef threats, dependence on reef ecosystem services (that is, social and economic sensitivity to reef loss), and the capacity to adapt to the potential impacts of reef loss.[217] As in the threat modeling, we use an indicator-based approach (Table 6.1). Exposure, which is based on modeled threats to reefs,[218] is combined with indicators of reef dependence and adaptive capacity to form an index of social and economic vulnerability to reef loss. This chapter examines aspects and global patterns of reef dependence, adaptive capacity, and overall vulnerability. For places identified as being most vulnerable, we consider the implications and underlying drivers of this vulnerability in detail, as a step toward more effectively targeting resources and efforts for management and development in reef-dependent regions.

REEF DEPENDENCE

Between tens and hundreds of millions of people worldwide rely on reef resources,[213, 219] depending on the types of stakeholders and resources considered. Global estimates of the economic values attributed to reef ecosystem services, although similarly coarse, range from tens to hundreds of billions of dollars (Box 6.3). Yet these numbers provide only a broad overview of the importance of reefs to economies, livelihoods, and cultures. Dependence on reefs is complex and highly variable among nations, communities, and individuals.[220] People may rely on reefs for one or multiple services, and this dependence can last year-round, during specific seasons, or only at critical times of hardship.[213] Reef dependence can also change over time, in response to large-scale drivers of change or alongside other cultural shifts.[221]

To capture the multidimensional nature of people's reliance on reefs, we break down reef dependence into six indicators that are important at the national scale: reef-associ-

BOX 6.1. VULNERABILITY ASSESSMENT METHODS

The three components of vulnerability to degradation and loss of reefs are outlined in Table 6.1, with the indicators used to assess them. We focused mainly at the national level because indicators of reef dependence and adaptive capacity are not available at finer scales for most reef nations. In total, 108 countries, territories, and subnational regions (e.g., states) are included in the study.[a] We obtained data from international organizations, published and grey literature, national statistics, and consultation with government officials and other experts. Where data were unavailable, we interpolated values based on countries or territories within the same region that were culturally and economically similar. Variables were normalized (i.e., rescaled from 0 to 1) and averaged when indicators comprised more than one variable. Indicators were then normalized and averaged to yield the three index components (exposure, reef dependence, and adaptive capacity). In turn, the components were normalized and multiplied together to yield the index of vulnerability.[222] We calculated all averages using equal weightings for each indicator. Results for components and vulnerability are presented as quartiles, with 27 countries and territories classified in each of four categories (low, medium, high, and very high). Technical notes, with full details of indicators, data sources, and methodology are available at www.wri.org/reefs.

a. The study includes 81 countries, 21 island territories, and six subnational regions (Florida, Hawaii, Hong Kong SAR, Peninsular Malaysia, Sabah, and Sarawak) that could be assessed separately because sufficient data were available. For simplicity, we refer to these as "countries and territories" throughout this chapter.

TABLE 6.1 VULNERABILITY ANALYSIS COMPONENTS, INDICATORS, AND VARIABLES

Component	Indicator	Variable
Exposure	Threats to coral reefs	• *Reefs at Risk* integrated local threat index weighted by ratio of reef area to land area
Reef dependence	Reef-associated population	• Number of coastal people within 30 km of reefs • Coastal people within 30 km of reefs as a proportion of national population
	Reef fisheries employment	• Number of reef fishers • Reef fishers as a proportion of national population
	Reef-associated exports	• Value of reef-associated exports as a proportion of total export value
	Nutritional dependence on fish and seafood	• Per capita annual consumption of fish and seafood
	Reef-associated tourism	• Ratio of registered dive shops to annual tourist arrivals, scaled by annual tourist receipts as a proportion of GDP
	Shoreline protection	• Index of coastal protection by reefs (combining coastline within proximity of reefs, and reef distance from shore)
Adaptive Capacity	Economic resources	• Gross domestic product (GDP) + remittances (payments received from migrant workers abroad) per capita
	Education	• Adult literacy rate • Combined ratio of enrollment in primary, secondary, and tertiary education
	Health	• Average life expectancy
	Governance	• Average of worldwide governance indicators (World Bank) • Fisheries subsidies that encourage resource conservation and management, as a proportion of fisheries value
	Access to markets	• Proportion of population within 25 km of market centers (> 5000 people)
	Agricultural resources	• Agricultural land area per agricultural worker

ated population, fisheries employment, nutritional dependence, export value, tourism, and shoreline protection (Box 6.1).

Indicators

Reef-associated population

Worldwide, roughly 850 million people live within 100 km of coral reefs, and are likely to derive some benefits from the ecosystem services they provide.[223] More than 275 million people reside within 10 km of the coast and 30 km of reefs, and these people are most likely to depend on reefs and reef resources for their livelihoods and well-being. Southeast Asia alone accounts for 53 percent of these most closely "reef-associated" people. For this assessment, we consider absolute (number of people) and relative (proportion of total population) measures of reef-associated population as coarse indicators of where people are most likely to rely on reefs.

In absolute numbers, countries like Indonesia and the Philippines have very large reef-associated populations, with tens of millions of coastal people living within 30 km of reefs (Figure 6.1). These large coastal populations increase both pressures on reefs and the likelihood that reef losses will have far-reaching social and economic consequences. In other nations—especially small-island states—absolute numbers of coastal people in close proximity to reefs may be smaller, yet represent a significant proportion of total population. In 40 countries and territories in our study, the entire population lives within 30 km of reefs; all of these—such as Anguilla, Kiribati, and Mayotte— are small islands. Here, although reef-associated populations are smaller, the relative impacts of reef loss may nevertheless be considerable.

FIGURE 6.1. COUNTRIES WITH THE LARGEST REEF-ASSOCIATED POPULATIONS

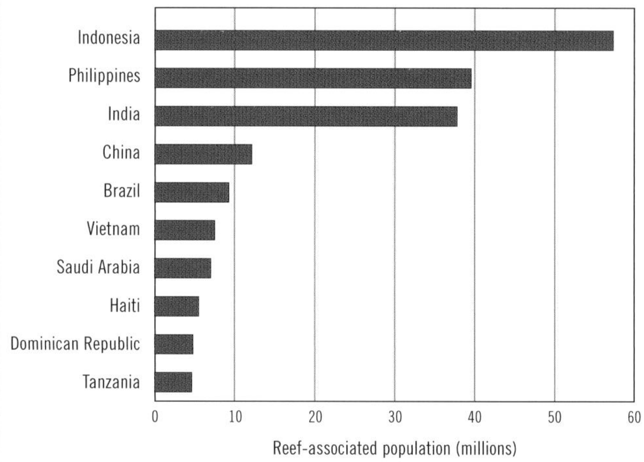

Note: Reef-associated populations are defined as those people living within 10 km of the coast, and within 30 km of reefs.

Fisheries employment

Fisheries are one of most direct forms of human dependence on reefs, providing vital food, income, and employment. They also play an important role in poverty alleviation. Reef fisheries are largely small-scale and artisanal, and many are open-access systems with relatively low entry costs, making them particularly attractive to poor and migrant people.[213] Inshore reefs are accessible even without fishing gear, and gleaning (harvesting by hand) is an important activity that is often predominantly carried out by women and children.[220, 224, 225] Although statistics from most reef nations have tended to underestimate the importance of reef fisheries,[16, 68] data on employment in reef fisheries have increasingly become available through recent large-scale socioeconomic studies in tropical coastal regions.[220, 226] We use two measures of employment for this assessment: (1) absolute numbers of people involved in reef fisheries, and (2) the relative proportion of total population that these fishers represent.[227]

Populous Asian nations account for the greatest absolute numbers of people who fish on reefs. Reef fishers in each of Indonesia, Philippines, India, Vietnam, and China are estimated to number between 100,000 and more than 1 million. Also within this range are two Pacific nations (New Caledonia and Papua New Guinea) and Brazil. As a proportion of total population, two-thirds of countries and territories with very high participation in reef fisheries[228] are in the Pacific (for example, Tokelau and Cook Islands), where the regional average of reef fisheries participation is 14 percent. The highest relative involvement in reef fishing (40 percent of the population) is reported from New Caledonia. The Turks and Caicos Islands, the Maldives, and Dominica are also among nations with significant proportions of reef fishers (5 to 7 percent of the population).

Reef-derived nutrition

Healthy reefs provide an abundant variety of foods, including fish, crustaceans, mollusks, sea cucumbers, and seaweeds. Many reef-derived foods represent inexpensive sources of high-quality animal protein. In some places—particularly small, isolated islands with limited resources and trade—they may be the only such source. Despite the critical importance of food from reefs, information about their consumption is limited. We therefore use estimates of fish and seafood consumption per capita[229] as the best available proxy measure of nutritional dependence on reefs, because they nevertheless provide a coarse indication of the importance of fish and seafood in diets.[230] Across reef nations and territories, people consume an average of 29 kg of fish and seafood each year. Consumption is highest in the Maldives (180 kg/person; Figure 6.2), where fish provide 77 percent of dietary animal protein.[184] The remaining nine of the top ten consumers are island countries and territories in the Pacific, a region where average consumption (57 kg/person) is nearly twice the global average, and concerns have been raised about potential shortfalls in fish supply by 2030.[231] Other places with very high consumption include Japan, Seychelles, Montserrat, Nauru, Malaysia, Fiji, and Antigua and Barbuda.

Reef-derived exports

Exports of reef-derived species and products represent important sources of revenue for tropical economies. Exports include many species and products from live and dead fish and invertebrates, and seaweeds. Some of these commodities are relatively high-value specialty items. For example, live reef fish imported for food in Hong Kong in 2008 were reportedly worth an average of nearly US$10/kg, while humphead wrasse, the most valuable species, were worth more than US$50/kg.[232] Reef exports may also be important regionally. In the Caribbean, spiny lobsters are the primary fishery for 24 countries, and represent a major source of export income for the region.[233]

FIGURE 6.2. CORAL REEF COUNTRIES AND TERRITORIES WITH THE HIGHEST FISH AND SEAFOOD CONSUMPTION

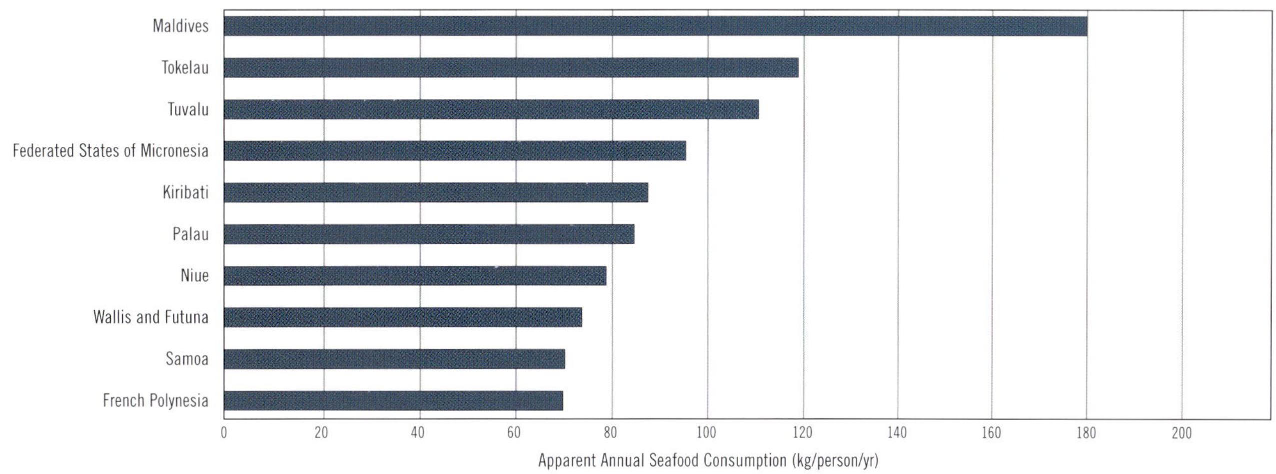

Note: Consumption includes marine and freshwater fish and invertebrates.

PHOTO: AMOS NACHOUM

Few countries specifically report the value of reef-derived exports,[234] and so we focus on the value of reef goods for which other sources of trade data are available: aquarium fish and invertebrates, sea cucumbers, black pearls, conch, corals, giant clams, live reef food fish, lobsters, seaweeds, and trochus (top shells).[235] Exports of these goods are reported by 96 countries and territories. In 21 countries and territories, reef-associated exports are valued at more than 1 percent of total exports, and in six cases, at more than 15 percent of total exports. The relative value of these exports is greatest for French Polynesia, where exports of black pearls (the territory's primary export commodity), trochus, aquarium fish, sea cucumbers, and corals are valued at 62 percent of national GDP. Other top exporters include the Turks and Caicos, Cook Islands, and the Bahamas.

Reef tourism

At least 96 countries and territories benefit from some level of reef tourism,[236] and in 23 countries and territories, reef tourism accounts for more than 15 percent of GDP.[237] Spending by divers and snorkelers supports a range of businesses (such as dive shops, hotels, restaurants, and transportation) and in some places directly contributes to the management costs of MPAs through visitor user fees. Other reef tourists include recreational fishers (for example, in Australia, Bahamas, and Cuba), and less directly, beach visitors, in areas where sand is supplied by nearby reefs.

For this analysis, we focus on the tourism sector that depends most directly on reefs: scuba diving. To capture the importance of dive tourism, we derived an indicator that combines the number of registered dive centers and the relative economic importance of tourism.[238] Relative to the number of tourists, the greatest numbers of dive shops are found in French and Dutch overseas territories, the Federated States of Micronesia, Solomon Islands, and the Marshall Islands. When dive shop numbers are considered in relation to their potential economic importance, the strongest dependence on dive tourism is found in Bonaire. This island is rated among the world's top diving destinations and more than half of its 74,000 visitors in 2007 were divers.[239] Other nations and territories that rely heavily on reef tourism are scattered across the Pacific (e.g., Palau), the Indian Ocean (e.g., Maldives), and the Caribbean (e.g., Belize).

Shoreline protection

Coral reefs play a valuable role in buffering coastal communities and infrastructure from the physical impacts of wave action and storms, thereby reducing coastal erosion and lessening wave-induced flooding. Coral reefs typically mitigate 75 to 95 percent of wave energy,[240] but are less effective for large waves or storm surges during storm events. The degree of shoreline protection provided by reefs varies with the coastal context. Important factors include the nature of the land protected (e.g., geology, vulnerability to erosion, and slope), the nature of the coral reef (depth, continuity, and distance from shore), and local storm and wave regimes.

To estimate the coastal protection that reefs provide, we derive an index of coastal protection based on the amount of coastline within proximity of reefs, with reefs closer inshore offering more protection than offshore reefs.[241] Globally, we estimate that more than 150,000 km of shoreline in 106 countries and territories receive some protection from reefs,[242] with an average of 42 percent protection at the national level. In 17 small islands (for example, Curaçao), more than 80 percent of the coastline is estimated to be protected by reefs. Reefs protected less than 10 percent of coastline in 21 countries and territories, which predominantly are large nations with relatively restricted areas of reef such as South Africa.

Results

Combining all six indicators reveals several clusters of particularly strong dependence on reefs (Map 6.1). More than half of the countries and territories with very high reef

MAP 6.1. SOCIAL AND ECONOMIC DEPENDENCE ON CORAL REEFS

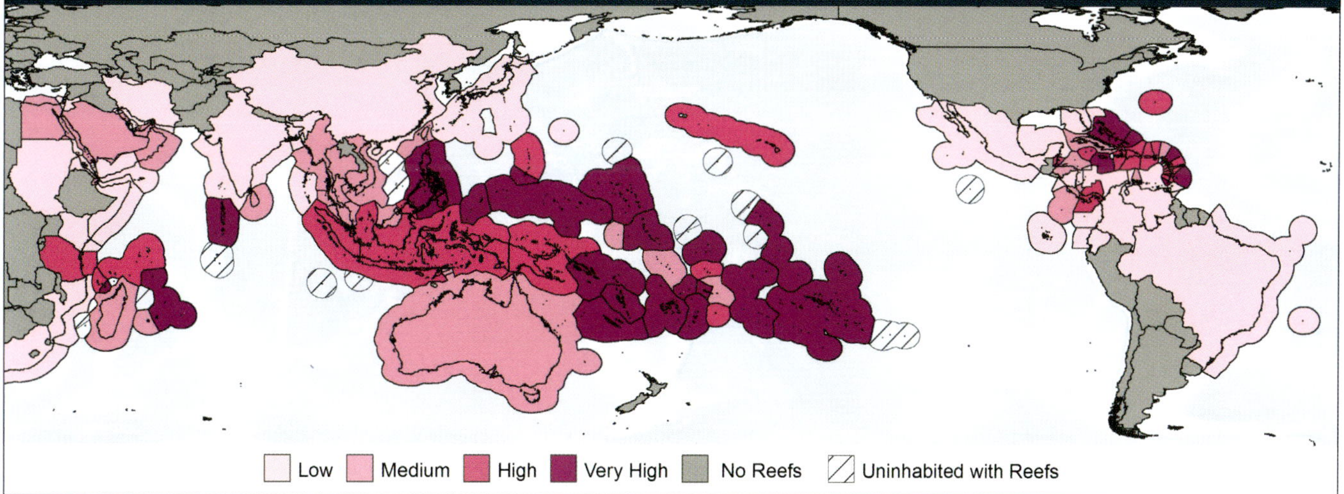

Notes: Reef dependence is based on reef-associated population, reef fisheries employment, nutritional dependence on fish and seafood, reef-associated export value, reef tourism, and shoreline protection from reefs. Eighty-one countries, 21 island territories, and six subnational regions (Florida, Hawaii, Hong Kong SAR, Peninsular Malaysia, Sabah, and Sarawak) were assessed, and are categorized according to quartiles. Reef territories that are only inhabited by military or scientific personnel are not included.

dependence are located in the Pacific (Table 6.2), including French Polynesia, which globally has the highest dependence on reefs. This ranking reflects the territory's predominantly reef-associated population, valuable reef-associated exports and tourism, significant consumption of fish and seafood, and extensive coastal protection from reefs. One-third of very high reef-dependent countries and territories are in the Caribbean, including Grenada, Curaçao, and the Bahamas. Nearly all the most strongly reef-dependent nations are small-island states. The only exception is the Philippines, which, with more than 7,000 islands and local jurisdiction over 22,500 sq km of coral reefs, could be argued to operate as a nation of many small-island states with respect to its reef resources.

All of the countries and territories that are very highly dependent on reefs are considered highly or very highly dependent on at least four separate indicators of reef dependence. In ten cases, dependence is high or very high on all six indicators: the Cook Islands, Fiji, Jamaica, the Maldives, the Marshall Islands, New Caledonia, the Philippines, Solomon Islands, Samoa, and Tonga. The Maldives are rated very high on all six measures of reef dependence. This atoll nation supports a growing tourism industry, relies upon reef fish for food (for residents and tourists), and also uses fish from reefs as live bait for catching valuable tuna.[243]

Reef-dependence is lowest where reefs make up only a small proportion of the coastline, or in very large countries. Even in these nations, some people and places may still rely strongly on particular reef ecosystem services, exemplified by the critical fishing and tourism sectors in places such as southern Florida in the United States, the Ryukyu Islands of southern Japan, and the Yucatan Peninsula in Mexico, and the strong nutritional reliance on reef fish among communities of the Lakshadweep Islands of India.[225]

ADAPTIVE CAPACITY

Adaptive capacity is the ability to cope with, adapt to, or recover from the effects of changes.[244] For nations faced with reef degradation and loss, adaptive capacity includes the resources, skills, and tools available for planning and responding to the effects of these losses (that is, the decreased flow of benefits from reef ecosystem services). Like reef dependence, adaptive capacity is complex and cannot be directly measured. We therefore separate adaptive capacity into six national-scale indicators that are relevant to reef-dependent regions (Table 6.1). We use two types of indicators: (1) those that describe general aspects of human and economic development, and (2) those that are more specific to the context of reef-dependent nations.[245]

MAP 6.2. CAPACITY OF REEF COUNTRIES AND TERRITORIES TO ADAPT TO REEF DEGRADATION AND LOSS

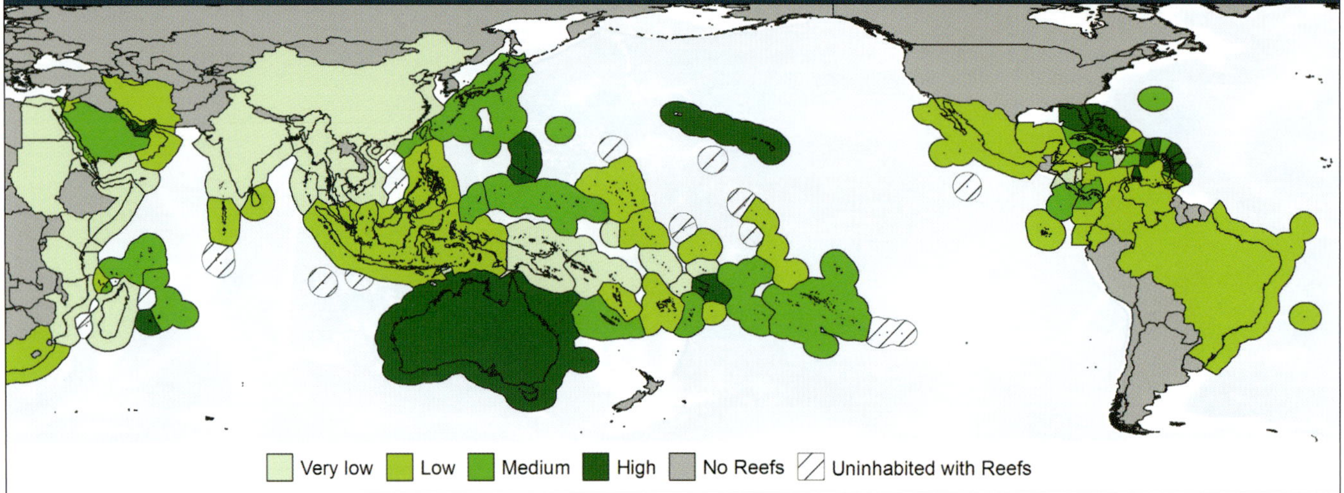

Notes: Adaptive capacity is based on economic resources, education, health, governance, access to markets, and agricultural resources. Eighty-one countries, 21 island territories, and six subnational regions (Florida, Hawaii, Hong Kong SAR, Peninsular Malaysia, Sabah, and Sarawak) were assessed, and are categorized according to quartiles.

In the face of potential losses of reef ecosystem services, we assume that adaptive capacity is likely to be stronger in countries and territories with healthier and more skilled populations, greater economic resources, and stronger governance. We use broad indicators of health (average life expectancy) and education (enrollment ratio in schools and adult literacy). To measure economic strength, we use per capita values of GDP[246] and remittances (payments received from migrant workers abroad). Remittances provide a vital source of income for some reef nations;[247] for example, Somalia and Tonga each receive remittances equal to more than 38 percent of their GDP. To describe the capacity of nations and territories to govern effectively, we include two measures: a composite index based on the World Bank's worldwide governance indicators,[248] and the value of subsidies that encourage more sustainable fisheries (for example, research, development, and MPA management).[249,250]

Two indicators capture aspects of adaptive capacity that are more specific to the case of potential reef loss. Access to markets is included as an indicator of isolation, in that reef-associated populations closer to market centers are likely to have more options for trading food and other goods in the event of reef loss. We use area of agricultural land as a proxy indicator of other natural resources to which reef-dependent people may turn if reefs are degraded or lost.[251] Many people in coastal communities engage in multiple livelihood activities,[252] and agriculture and fishing are frequently carried out in parallel. Where reef ecosystem services decline, the agricultural sector is likely to be placed under additional demands for food production and employment.

Results

When these six indicators are combined, we find that adaptive capacity is most limited for nations with a relatively recent history of conflict, such as Somalia, Mozambique, Eritrea, Sudan, and Timor-Leste (Table 6.2; Map 6.2). Most of the reef countries classified as LDCs (15 of 19) fall within the lowest category of adaptive capacity, including the five countries listed above and others such as Bangladesh, Tanzania, and Yemen.[215] Not surprisingly, adaptive capacity is typically greatest among countries characterized by high levels of economic development and resources (for example, the United States and Singapore), including oil-producing nations (such as Brunei and Qatar) and Caribbean islands engaged in offshore finance (such as the British Virgin Islands and Cayman Islands).

SOCIAL AND ECONOMIC VULNERABILITY

Combining the three components of vulnerability reveals that the countries and territories that are most vulnerable to the degradation and loss of reefs are spread throughout the world's tropical regions (Map 6.3). More than one-third of

TABLE 6.2. COUNTRIES AND TERRITORIES WITH HIGHEST THREAT EXPOSURE, STRONGEST REEF DEPENDENCE, AND LOWEST ADAPTIVE CAPACITY

Highest exposure to reef threats	Highest reef-dependence	Lowest adaptive capacity
American Samoa	Antigua and Barbuda	Bangladesh
Anguilla	Bahamas	Cambodia
Antigua and Barbuda	Barbados	China
Aruba	Cook Islands	Djibouti
Bahrain	Curaçao	Egypt
Barbados	Federated States of Micronesia	Eritrea
Bermuda	Fiji	Haiti
British Virgin Islands	French Polynesia	India
Comoros	Grenada	Kenya
Curaçao	Guam	Madagascar
Dominica	Jamaica	Montserrat
Dominican Republic	Kiribati	Mozambique
Grenada	Maldives	Myanmar
Guadeloupe	Marshall Islands	Nauru
Haiti	Mauritius	Nicaragua
Jamaica	Mayotte	Papua New Guinea
Martinique	New Caledonia	Solomon Islands
Mayotte	Palau	Somalia
Nauru	Philippines	Sudan
Northern Mariana Islands	Samoa	Tanzania
Philippines	Solomon Islands	Thailand
Puerto Rico	St. Kitts and Nevis	Timor-Leste
Samoa	St. Lucia	Tokelau
St. Eustatius	Tonga	Tuvalu
St. Kitts and Nevis	Turks and Caicos	Vietnam
St. Lucia	Vanuatu	Wallis and Futuna
Virgin Islands (U.S.)	Wallis and Futuna	Yemen

Notes: For each component, the 27 countries and territories in the highest quartile of risk are shown: very high exposure to reef threats, very high reef dependence, and low adaptive capacity. Countries and territories are listed alphabetically.

MAP 6.3. SOCIAL AND ECONOMIC VULNERABILITY OF COUNTRIES AND TERRITORIES TO REEF LOSS

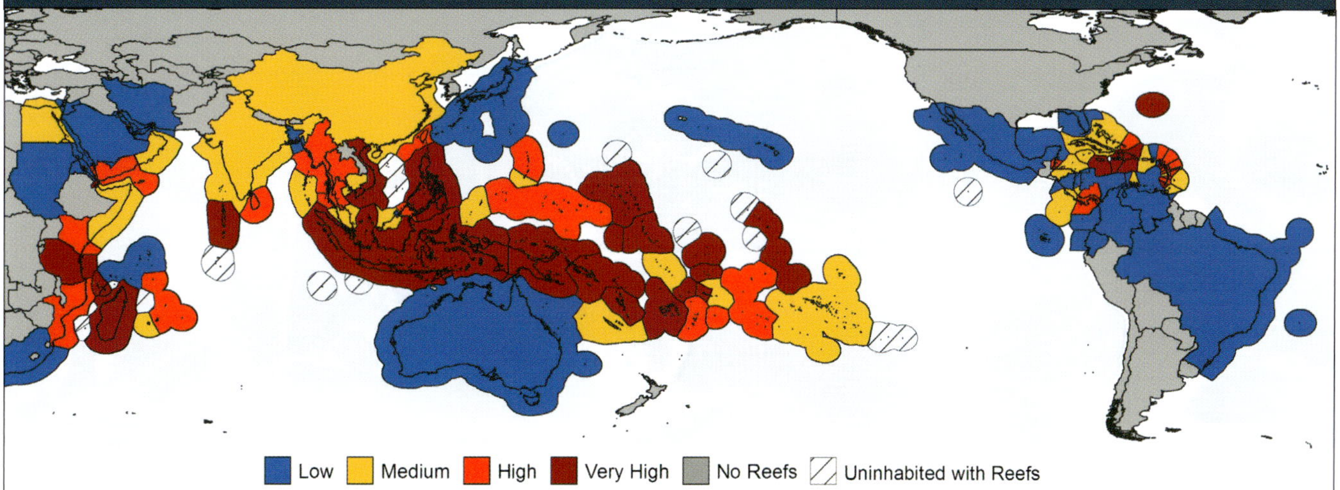

Notes: Vulnerability is based on exposure to reef threats, reef-dependence, and adaptive capacity. Eighty-one countries, 21 island territories, and six subnational regions (Florida, Hawaii, Hong Kong SAR, Peninsular Malaysia, Sabah, and Sarawak) were assessed, and are categorized according to quartiles.

FIGURE 6.3. DRIVERS OF VULNERABILITY IN VERY HIGHLY VULNERABLE NATIONS AND TERRITORIES

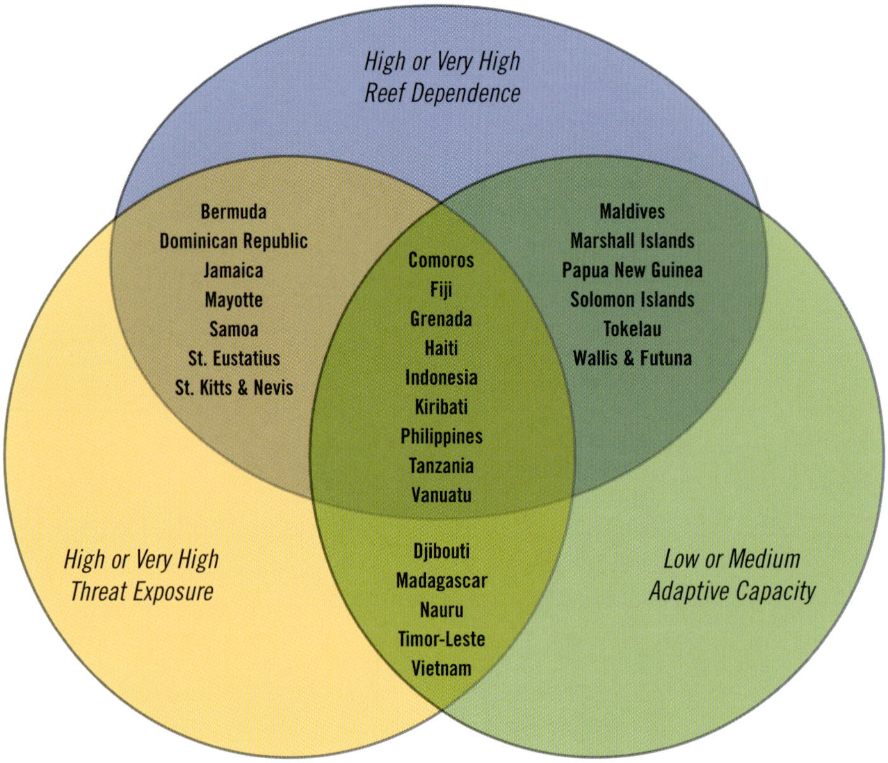

Note: Only the 27 very highly vulnerable countries and territories are shown.

very highly vulnerable countries and territories are in the Caribbean, one-fifth are in Eastern Africa and the Western Indian Ocean, and smaller numbers are found in the Pacific, Southeast Asia, and South Asia. Among the 27 countries and territories rated as very highly vulnerable, the majority (19) are small-island states.

The most vulnerable countries and territories reflect different underlying combinations of the three components (exposure, reef dependence, and adaptive capacity) (Figure 6.3). Each of these types of vulnerability has different implications for the likely consequences of reef loss; identifying them provides a useful starting point for setting priorities for resource management and development action to minimize potential impacts. It may also provide an opportunity for countries that are not considered highly vulnerable to plan how best to avoid future potential pitfalls.

Nine countries (Comoros, Fiji, Grenada, Haiti, Indonesia, Kiribati, Philippines, Tanzania, and Vanuatu) lie in a position of serious immediate social and economic vulnerability, with high to very high exposure and reef dependence, and low to medium adaptive capacity. These nations represent key priorities for concerted national and local efforts to reduce reef dependence and build adaptive capacity, alongside reducing immediate threats to reefs. These efforts should ideally be integrated within the broader national development context, and where possible, within other ongoing development initiatives. For example, Haiti, the most vulnerable country in the study and the poorest nation in the Western Hemisphere, is currently engaged in rebuilding lives, livelihoods, and infrastructure following a devastating earthquake in January 2010. Recognizing the needs of reef-dependent communities within such efforts may bring opportunities for reducing their vulnerability to future reef loss, as well as identifying the role that sustainable use of reef resources can play in poverty reduction and economic development.

For six island countries and territories (the Maldives, the Marshall Islands, Papua New Guinea, Solomon Islands,

Tokelau, and Wallis and Futuna), where exposure to reef threats is not yet extreme at the national scale, strong reliance on reefs and limited capacity to adapt suggest that if pressures on reefs increase, serious social and economic impacts may result. This situation may offer a window of opportunity to build secure management frameworks to protect reefs, shift some human dependence away from reefs, and strengthen local and national capacity. The window may be limited, however, given that large-scale threats such as climate change (which is not included within the exposure index) may also have serious consequences on reefs. Such large-scale impacts have already occurred in some of these places. For example, the Maldives experienced severe losses of coral following bleaching in 1998,[162] while reefs in the western Solomon Islands were affected by an earthquake and tsunami in 2007, with resulting impacts on coastal communities and fisheries.[253]

Seven very highly vulnerable countries and territories in the Caribbean, Western Indian Ocean, and Pacific (Bermuda, the Dominican Republic, Jamaica, Mayotte, Samoa, St. Eustatius, and St. Kitts and Nevis) have reefs that are highly or very highly exposed to threat and depend heavily on reef ecosystem services, but also have high or very high levels of adaptive capacity.[254] All of these are small-island states, and many are densely populated. While relatively high adaptive capacities are likely to help these islands to buffer potential impacts on reef-dependent people, ultimately the extent of their vulnerability to reef loss will depend on how effectively resources and skills are directed toward reducing reef threats and dependence. For example, Jamaica is considered to be an upper-middle-income country,[255] and derives earnings from agriculture, mining, and tourism, among other sources. However, more than three-quarters of households in some communities depend on fishing for their livelihoods.[220] Years of heavy fishing pressure have already contributed to marked declines in the country's reefs,[256] further adding to the need to develop feasible economic alternatives for fishers.

In five reef nations (Djibouti, Madagascar, Nauru, Timor-Leste, Vietnam), very high vulnerability stems from serious threats to reefs and limited adaptive capacity, despite only moderate national-scale dependence upon reefs. This combination of drivers suggests that while social and economic impacts of reef loss may be serious for some local areas, these effects are likely to be less significant on a national scale. In these countries, vulnerability may be reduced most effectively by targeting efforts to reduce threats to reefs and build capacity at local scales and by raising awareness within relevant government agencies about regions where reef dependence is particularly high. Attention should also be paid to cases where reef-dependence may increase. For example, although most of the population of Nauru is employed by the government, recent economic hardship on the island has increased direct dependence on coastal resources, including reefs, for food.[257]

BOX 6.2 REEF STORY

Philippines: Social Programs Reduce Pressure on Culion Island's Reefs

Culion Island, in the southwestern Philippines, is surrounded by diverse reefs. In coastal villages, rapid growth in population, heavy dependence on coastal resources, and destructive fishing practices have resulted in the near collapse of reef habitat and fisheries. To address these concerns, PATH Foundation Philippines started the Integrated Population and Coastal Resource Management (IPOPCORM) initiative to empower communities to implement family planning activities simultaneously with community-led coastal conservation and alternative livelihood strategies. This approach has led to increased community well-being, greater food security, and an improvement in the health of Culion's reefs. *See full story online at www.wri.org/reefs/stories.*

Story provided by Leona D'Agnes, Francis Magbanua, and Joan Castro of the PATH Foundation Philippines, Inc.

PHOTO: PATH FOUNDATION

LIMITATIONS AND CHALLENGES OF THE ANALYSIS

This study represents the first global assessment of the vulnerability of countries and territories to reef loss, and several limitations apply to the analysis. First, we compiled data primarily from published sources, and could not include some potentially relevant indicators of reef dependence (for example, employment in fish processing and trade) or adaptive capacity (for example, perception of risk and capacity for self-organization) because data were unavailable for most countries and territories. Second, we focused mainly at the national level, due to the global scope of the study and data limitations. Country-level assessments such as this are useful for providing a broad view of vulnerability patterns, but they do not reveal the distribution of vulnerability within countries. At the national level, average vulnerability may be low, yet pockets of high vulnerability may still exist, reflecting local variation in reef threats, reef dependence, and adaptive capacity. For example, India was rated as a medium vulnerability country, with low reef-dependence, but a finer-scale assessment would likely find higher reef-dependence, and as a result, greater vulnerability, in specific regions such as India's Andaman and Nicobar Islands. Third, while we have made every effort to gather data from standardized or comparable sources, some variation in the quality of estimates among countries is unavoidable. For example, data on numbers of fishers (particularly subsistence fishers) are likely to be underestimated in some regions. Finally, while we focus on vulnerability to the loss of reef ecosystem services, other factors also represent significant threats to reef nations, but are beyond the scope of this assessment. For example, sea level rise also poses a considerable danger to coastal communities, infrastructure, and livelihoods, particularly in low-lying island nations.

CONCLUSIONS

Globally, the extent of reef dependence is enormous. Threats to reefs have the potential to bring significant hardship to many coastal communities and nations for whom livelihoods, food, and income are closely intertwined with reef ecosystem services. This vulnerability is greatest where high levels of reef threat coincide with heavy dependence on reefs and limited societal capacity to adapt or cope with reef loss. Ultimately, reducing vulnerability depends partly on management to reduce or eliminate local threats to reefs, but equally, requires measures to shift dependence, at least partially, away from reefs. This need will become even greater as climate change impacts on reef ecosystems become more frequent and severe. Reef-dependent people may be vulnerable to climate-driven reef losses, even where local threats to reefs are minimal. Yet efforts to reduce reef dependence are extremely challenging. Planning and prioritizing at local scales are hindered by a lack of information about dependence on specific reef ecosystem services (for example, dietary consumption, numbers of subsistence fishers) in many areas. Even where reef dependence is well-understood, past efforts to develop alternative livelihoods in coastal areas have frequently proven unsuccessful.[258] Activities such as agriculture, aquaculture, tourism, or trade may represent viable alternatives, but will only be sustainable where their development takes into account local aspirations, needs, perceptions, and cultural ties to coral reefs.[259] For millions of reef-dependent people, it is critical that such efforts succeed.

PHOTO: AMY V. UHRIN

BOX 6.3. ECONOMIC VALUE OF CORAL REEFS

Economic valuation is a tool that can aid decisionmaking by quantifying ecosystem services, such as those provided by coral reefs, in monetary terms. In traditional markets, ecosystem services are often overlooked or unaccounted for, an omission that regularly leads to decisions favoring short-term economic gains at the expense of longer-term benefits. Economic valuation provides more complete information on the economic consequences of decisions that lead to degradation and loss of natural resources, as well as the short- and long-term costs and benefits of environmental protection.

CORAL REEF VALUES

Many studies have quantified the value of one or more ecosystem services provided by coral reefs. These studies vary widely in terms of spatial scale (from global to local), method used, and type of value estimated. Some assessments focus on the annual benefits coming from reefs, and some estimate total value over a number of years. Still others focus on the change in value as an ecosystem is altered (such as the reduction in shoreline protection due to the degradation of a coral reef). Of the many ecosystem services provided by coral reefs, reef-related fisheries, tourism, and shoreline protection are among the most widely studied because their prices are traceable in markets and are thus relatively easy to calculate. Although cultural, aesthetic, and future benefits associated with reefs are also significant, they have largely been absent in valuation studies due to a lack of information from existing or comparable markets.[260] We describe the valuation methods for reef-related fisheries, tourism, and shoreline protection services below, and provide examples of values in Table 6.3. The economic benefits derived from coral reefs vary considerably by site, depending on the size of tourism markets, the importance and productivity of fisheries, level of coastal development, and the distance to major population centers. Estimating such values is not easy; some of the challenges and limitations of economic valuation are described below.

Tourism. Estimating the economic value of coral reef-associated tourism typically focuses on its contributions to the economy, through tourist expenditures, adjusted for the operating costs of providing the service. A recent summary of 29 published studies on reef-associated tourism found a very wide range in values, from about US$2/ha/yr to US$1 million/ha/yr.[261] However, most values fall within the narrower range of US$50/ha/yr to US$1,000/ha/yr. The wide variation of values is strongly related to differences in the accessibility of places, with very low tourism values in remote locations that have limited tourism development and very high values in areas that have intensive tourism. For these reasons, it is not possible to undertake simple extrapolations of specific studies to entire reef tracts where demand and access may be very different.

In more than 20 of the 100 reef countries and territories that benefit from reef-associated tourism, tourism accounts for more than 30 percent of export earnings.[8]

Fisheries. Valuation of coral reef-associated fisheries typically focuses on their economic contribution based on landings, sales, market prices, and operating costs. The productivity of a reef fishery and its sustainable yield should also be considered, although assessing values for both of these can be challenging due to ongoing fishing pressure and habitat degradation. In many areas, the potential fishery value could be considerably higher than its current value if fisheries were better managed and fish habitat and stocks recovered.[262]

A healthy, well managed reef can yield between 5 and 15 tons of seafood per sq km per year.[6, 7]

Shoreline Protection. Estimates of shoreline protection value are typically based on anticipated property losses under different levels of reef degradation, and are therefore highly dependent on property values. The value of shoreline protection services from reefs is usually estimated through one of two methods. "Replacement cost" estimates the cost of providing a substitute for shoreline protection by reefs with alternative manmade structures. "Avoided damages," on the other hand, estimates the reduction in inundation and damage due to the presence of the reef, and couples this with the current value of assets and property protected to determine the value.

Over 150,000 km of shoreline in 100 countries receive at least some protection from coral reefs.[263]

continued

BOX 6.3. *continued*

TABLE 6.3. SAMPLE VALUES: ANNUAL NET BENEFITS FROM CORAL REEF-RELATED GOODS AND SERVICES (US$, 2010)

Extent of Study	Tourism	Coral-reef Fisheries	Shoreline Protection
Global [a]	$11.5 billion	$6.8 billion	$10.7 billion
Caribbean (Regional)[b]	$2.7 billion	$395 million	$944 million to $2.8 billion
Philippines & Indonesia[c]	$258 million	$2.2 billion	$782 million
Belize (National)[d]	$143.1 million to $186.5 million**	$13.8 million to $14.8 million**	$127.2 to $190.8 million
Guam (National)[e]	$100.3 million**	$4.2 million**	$8.9 million
Hawaii (Subnational)[f]	$371.3 million	$3.0 million	Not evaluated

* All estimates have been converted to US$ 2010.

** Estimates of the value of coral reef-associated fisheries and tourism for Belize and Guam are gross values, while all other numbers in the table are net benefits, which take costs into account.

a. Cesar, H., L. Burke, and L. Pet-Soede.2003. *The Economics of Worldwide Coral Reef Degradation.* Zeist, Netherlands: Cesar Environmental Economics Consulting (CEEC).

b. Burke, L., and J. Maidens. 2004. *Reefs at Risk in the Caribbean.* Washington, DC: World Resource Institute.

c. Burke, L., E. Selig, and M. Spalding.2002. *Reefs at Risk in Southeast Asia.* Washington, DC: World Resources Institute.

d. Cooper, E., L. Burke, and N. Bood. 2008. *Coastal Capital: Belize The Economic contribution of Belize's coral reefs and mangroves.* Washington, DC: World Resource Institute.

e. Haider, W. et al. 2007. *The economic value of Guam's coral reefs.* Mangilao, Guam: University of Guam Marine Laboratory.

f. Cesar, H. 2002. The biodiversity benefits of coral reef ecosystems: Values and markets. Paris: OECD.

VALUATION OF LOSSES DUE TO DEGRADATION

Although many economic valuation studies have focused on estimating the benefits of coral reef ecosystem services, some studies have also focused on changes in value—that is, what an economy stands to lose if a reef is degraded. For example, the 2004 *Reefs at Risk in the Caribbean* study estimated that, by 2015, the projected degradation of Caribbean reefs from human activities such as overfishing and pollution could result in annual losses of US$95 million to US$140 million in net revenues from coral reef-associated fisheries, and US$100 million to US$300 million in reduced tourism revenue. In addition, degradation of reefs could lead to annual losses of US$140 million to US$420 million from reduced coastal protection within the next 50 years.[264] Other studies estimate that Australia's economy could lose US$2.2 billion to US$5.3 billion over the next 19 years due to global climate change degrading the Great Barrier Reef,[265] while Indonesia could lose US$1.9 billion over 20 years due to overfishing.[266]

POLICY AND MANAGEMENT APPLICATIONS

Ultimately, the goal of economic valuation is to influence decisions that will promote sustainable management of reefs. By quantifying the economic benefits or losses likely to occur due to degradation of reefs, it is possible to tap public and private funding for coastal management, gain access to new markets, initiate payments for ecosystem services, and charge polluters for damages. There are numerous examples of economic analyses successfully informing policy. For example, in the United States, the states of Hawaii and Florida adopted legislation setting amounts for monetary penalties per square meter of damaged coral reef, based on calculations from valuation studies.[267] The Belize government used an economic valuation study of its coral reefs as the premise to sue for damages after the container ship *Westerhaven* ran aground on its reef in January 2009, resulting in the Belizean Supreme Court ruling that the ship's owners must pay the government US$6 million in damages.[268] Finally, Bonaire National Marine Park, one of the world's few self-financed marine parks, used economic valuation to determine appropriate user fees.[269]

CHALLENGES AND LIMITATIONS

Despite the usefulness of economic valuation, there are still many challenges to its practical application. This is evidenced by the wide variation in the quality and consistency of existing economic valuation studies. In particular, although global-scale valuation studies are frequently cited, they are often misleading due to the difficulty of aggregating values and constraints on data at the global level. Furthermore, economic valuation can produce only a partial estimate of total ecosystem value, as humankind's limited technical, economic, and ecological knowledge prevents us from ever truly identifying, calculating, and ranking all of an ecosystem's values. Ultimately, valuation results should be used as part of a larger decisionmaking "toolbox" rather than being relied upon in a vacuum. In particular, valuation studies need to take into account the local context—both social and biological—and be undertaken with an eye toward the bigger picture. For example, what are the implications of under- or overestimating a given ecosystem service? Who will be the winners and losers if an ecosystem service is gained or lost? It is critically important when undertaking economic valuation to engage local stakeholders, document all data and assumptions, and carefully explain the uses and limitations of the research.

Chapter 7. SUSTAINING AND MANAGING CORAL REEFS FOR THE FUTURE

PHOTO: MARK SPALDING

Despite an overall picture of rising levels of stress and of failing health and productivity on many of the world's coral reefs, people *can* live sustainably alongside reefs, deriving considerable benefits from them. The challenges, as societies grow and technologies change, are to understand the limits to sustainability and to manage human activities to remain within these limits. How many fish can we take before we start to impact future food security and ecological stability? How can a village or a region fairly and effectively ensure access to fish without exceeding such limits? We now understand much more about the complex ecology of coral reefs and have developed a broad range of tools for reef management, but challenges remain in applying them.

This chapter focuses on the role of managed areas—notably marine protected areas (MPAs) and locally managed marine areas (LMMAs)—in protecting coral reefs. Such areas are the most widely used tools in coral reef management and conservation, and are the only tools for which sufficient data were available to conduct a global analysis. The chapter first briefly discusses reef management approaches in general, and then presents the first-ever global assessment of reef coverage in managed areas, including an assessment of their effectiveness.

REEF PROTECTION APPROACHES

Beyond marine managed areas, there exists a broad range of other management approaches that support reef health and resilience. Numerous fisheries management tools (regulations regarding fishing grounds, catch limits, gears, fishing seasons, or the capture of individual species) are often applied independently from MPAs. Other management measures deal with marine-based threats, for example through controls on discharge from ships, shipping lanes, and anchoring in sensitive areas. Land-based sources of sediment and pollution are managed through coastal zone planning and enforcement, sewage treatment, and integrated watershed management to reduce erosion and nutrient runoff from agriculture. These approaches are described in greater detail in Chapter 3, which presents remedies proposed for broad categories of reef threats, and in Chapter 8, which presents overall recommendations for reef conservation.

Communications are critical, both for improving understanding of risks, and ensuring sustained application of management measures; in many cases simply informing communities of alternative management approaches can lead to rapid changes. Incentives can also play an important role. Examples include training reef users to ensure sustainable practices, provision of alternative livelihoods, or even direct financial interventions such as payment for ecosystem services, where local communities—considered owners or stewards of an ecosystem—are paid in cash or kind for the benefits provided by the ecosystem.

Marine Protected Areas

MPAs are one of the most widely used management tools in reef conservation. Simply defined, an MPA is any marine area that is actively managed for conservation.[270] Such a definition is broad. At one end of the scale, it includes areas with just a few restrictions on fishing or other potentially harmful activities, even without a strict legal framework. At the other, it extends to sites with comprehensive protection targeting multiple activities, including recreational boating, fishing, pollution, and coastal development.

At their most effective, MPAs are able to maintain healthy coral reefs even while surrounding areas are degraded; support recovery of areas that may have been overfished or affected by other threats; and build resilient reef communities that can recover more quickly than non-protected sites from a variety of threats, including diseases and coral bleaching.[61, 62, 97, 173, 271] Of course, such areas are not immune from impacts. In most cases they offer only a proportional reduction in impacts, and degradation within MPAs is still a major problem.[81, 172, 272] The most consistent feature of MPAs is the provision of some control over fishing, although few offer complete protection. Many MPAs place other restrictions on activities such as boat anchoring, tourism use, or pollution. In addition, they are valuable for research, education, and raising awareness about the importance of an area. Where sites extend into adjacent terrestrial areas, they may provide additional benefits, such as limiting coastal development or other damaging types of land use. Even ineffective sites offer a basis on which future, more effective, management can be built.

MPAs are most successful when they have support from local communities.[273, 274] This is often achieved by involving local stakeholders in planning processes, which may include participation in activities such as site selection, resource assessment, and monitoring. Even where ownership and management of resources remains the domain of a government agency, education, consultation, and debate across the community lies at the heart of successful management. The Great Barrier Reef Marine Park Authority provides powerful evidence for this approach, with the successful expansion of closed zones across the park, as a result of years of consultation with local communities.[275]

In some countries, this process of stakeholder involvement has been extended to full local ownership and management.[273, 276, 277] Such an approach typically requires significant changes in governance structures. For example, in the Philippines, where reefs are among the world's most threatened, local municipalities have been given partial jurisdiction over inshore waters, including nearshore coral reefs. This has led to a burgeoning of new local fisheries regulations and MPAs.[278] When local communities understand that the benefits from such management only come from full compliance, they are also much more likely to police them vigorously.[279]

The effectiveness of individual MPAs can be greatly enhanced if they exist in a broad framework of protection covering wide areas or multiple sites. This may be achieved through very large MPAs (often zoned), or through the development of networks of sites that enable the maintenance of healthy reef populations at multiple locations. Such large-scale approaches provide some security against impacts or losses at individual sites and support the movement of adults and of eggs and larvae between locations.[280] Applying both social and ecological knowledge to the development of MPA systems or networks can increase such benefits, both through incorporating human needs and pressures, and by ensuring that biodiversity is fully covered, and that natural movements can be maintained or maximized.[281]

Locally Managed Marine Areas

The trend toward ownership of marine space or resources at local levels has led, in many areas, to more comprehensive management strategies. Locally managed marine areas (LMMAs) are marine areas that are "*largely or wholly managed at a local level by the coastal communities, landowning groups, partner organizations, and/or collaborative government representatives who reside or are based in the immediate area.*"[276] Under this definition, LMMAs are not areas set aside for conservation *per se*, but are managed for sustainable use. Most LMMAs restrict resource use, and many contain permanent, temporary, or seasonal fishery closures as well as other fisheries controls. In this way, LMMAs in their entirety are similar to many MPAs with no-take zones or wider areas of restricted use.

The best examples of LMMAs are in the Pacific region, where most reefs were held in customary tenure by adjacent villages for centuries. Such ownership was weakened or lost in some areas during the twentieth century, but recent decades have seen more formal legal recognition of traditional ownership in countries such as Fiji, the Solomon Islands, and Vanuatu.[7] In these places, local communities have begun to take management control over their marine resources so that they may be recognized as LMMAs. A further advantage of such local management is the rapid transmission of ideas between neighboring communities and islands; for example, there has been a rapid proliferation of small no-take reserves in LMMAs across parts of Vanuatu.[273, 274, 276, 277]

This application of local management is clearly important. Scaled-up across multiple locations and communities, LMMAs could prove as important for coral reef conservation as the designation of very large-scale MPAs in remote areas where local threats are minimal. For the sake of simplicity, references to MPAs for the remainder of this chapter also include LMMAs.

BOX 7.1 REEF STORY

Fiji: Local Management Yields Multiple Benefits at the Namena Marine Reserve

The Namena Marine Reserve surrounds the 1.6 km-long island of Namenalala and one of Fiji's most pristine reef ecosystems—the Namena Barrier Reef. In the mid-1980s, community members began noticing drastic declines in fish populations on the reef due to intensive commercial fishing. As a result, local chiefs and community leaders led a movement against commercial fishing that ultimately resulted in the establishment of a locally managed marine area (LMMA) network. Managing the LMMA emphasizes an ecosystem-based management approach while also protecting traditional fishing practices and creating tourism revenue to support a scholarship fund for local children. The reefs are recovering, providing an invaluable lesson in how community action combined with management knowledge can provide multiple benefits. *See full story online at www.wri.org/reefs/stories.*

Story provided by Stacy Jupiter of the Wildlife Conservation Society, Fiji and Heidi Williams of Coral Reef Alliance.

The global coverage of MPAs

There are an estimated 2,679 coral reef protected areas worldwide, encompassing approximately 27 percent of the world's coral reefs (Table 7.1).[282] There is considerable geographic variation in this coverage: while more than three-quarters of Australia's coral reefs are within MPAs, outside of Australia the area of protected reefs drops to only 17 percent.

TABLE 7.1 REGIONAL COVERAGE OF CORAL REEFS BY MPAs

REGION	No. of MPAs	Reef Area in MPAs (sq km)	Total Reef Area (sq km)	Reefs in MPAs (%)
Atlantic	617	7,630	25,850	30
Australia	171	31,650	42,310	75
Indian Ocean	330	6,060	31,540	19
Middle East	41	1,680	14,400	12
Pacific	921	8,690	65,970	13
Southeast Asia	599	11,650	69,640	17
Global Total	**2,679**	**67,350**	**249,710**	**27**

While these overall protection figures are high—few other marine or terrestrial habitats have more than one-quarter of their extent within protected areas—there is still cause for concern. First, most of the remaining 73 percent of coral reefs lie outside any formal management framework. Second, it is widely agreed that not all MPAs are effective in reducing human threats or impacts. Some sites, often described as "paper parks," are ineffective simply because the management framework is ignored or not

MAP 7.1. MARINE PROTECTED AREAS IN CORAL REEF REGIONS CLASSIFIED ACCORDING TO MANAGEMENT EFFECTIVENESS RATING

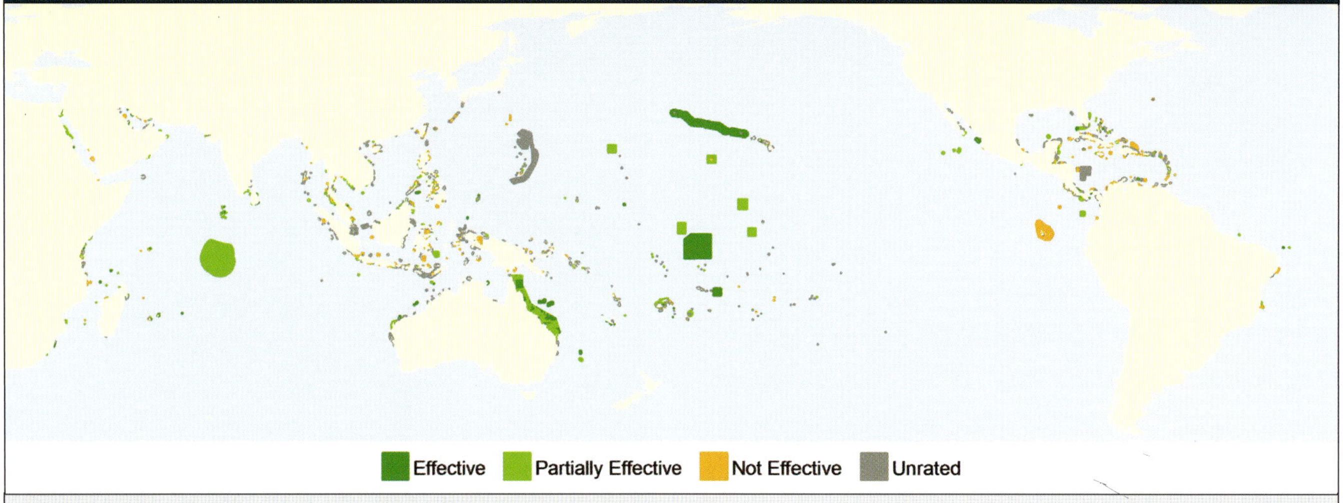

Notes: MPAs for coral reef regions were rated by regional experts according to their effectiveness level using a 3-point score. 1) MPAs rated as "effective" were managed sufficiently well that local threats were not undermining natural ecosystem function. 2) MPAs rated as "partially effective" were managed such that local threats were significantly lower than adjacent non-managed sites, but there may still be some detrimental effects on ecosystem function. 3)MPAs rated as "not effective" were unmanaged, or management was insufficient to reduce local threats in any meaningful way.

enforced. In others, the regulations, even if fully and effectively implemented, are insufficient to address the threats within its borders. For example, a site that forbids the use of lobster traps, but permits catching lobsters by hand may be just as thoroughly depleted of lobsters and suffer as much physical damage from divers as if the trap restrictions were not in place. Third, MPAs are rarely placed in areas where threats to reefs are greatest. This is highlighted by the recent creation of a number of very large MPAs in remote areas, where there are few or no local people and where threats are very low. Such MPAs are clearly important as potential regional strongholds, refuges, and seeding grounds for recovery, but do very little to mitigate current, urgent, local threats.

A further problem is that many reefs are affected by threats that originate far away, particularly pollutants and sediments from poor land-use practices or coastal development in areas outside the MPA boundaries. While healthy reefs may be more resilient to such stresses, this alone is unlikely to be sufficient, and other management approaches may be required to deal with these issues. In a few cases, considerable progress has been made through the engagement of adjacent communities to improve land management and reduce pollution and sediment runoff in areas adjacent to MPAs.[73]

MANAGEMENT EFFECTIVENESS AND CORAL REEFS

There is no single agreed-upon framework to assess how well MPAs reduce threats, although considerable resources are now available to support such assessments.[283] For this work, we undertook a rapid review—with a limited scope—to try to assess the effectiveness of MPA sites at reducing the threat of overfishing in as many sites as possible.[284] Our interest was to capture the ecological effectiveness of sites. Sites that were deemed ineffective or only partially effective could be scored in such a way because of the failure of implementation *or* because the regulatory and management regime allowed for some ecological impacts. We obtained scores from regional experts for 1,147 sites. These sites represent about 43 percent of our documented MPAs, but cover 83 percent of all reefs in MPAs by area (as we have scores for most of the larger MPAs). These results are summarized by region in Table 7.2. Figure 7.1 reflects management effectiveness ratings for all MPAs.

Our analysis revealed that nearly half (47 percent) of the 1,147 coral reef MPAs for which we have ratings are considered ineffective in reducing overfishing. Furthermore, the proportion of ineffective sites is highest in the most threatened regions of world: 61 percent of MPAs in the Atlantic and 69 percent of MPAs in Southeast Asia are rated as ineffective. Even these statistics are probably conservative,

TABLE 7.2 EFFECTIVENESS OF CORAL REEF-RELATED MARINE PROTECTED AREAS BY REGION

Region	No. of sites	Sites rated	Proportion of rated sites (%) Effective	Partial	Not effective
Atlantic	617	310	12	26	61
Australia	171	27	44	52	4
Indian Ocean	330	192	29	46	25
Middle East	41	27	33	37	30
Pacific	921	252	18	57	25
Southeast Asia	599	339	2	29	69
Global Total	**2,679**	**1,147**	**15**	**38**	**47**

as it is likely that our sampling favors better-known sites, many of which would have stronger management regimes than less well-known sites.

In comparing coral reef locations and management effectiveness, we find that, by area, 6 percent of the world's reefs are located in MPAs rated as effectively managed and 13 percent are located in areas rated as partially effective. Four percent of reefs are in areas rated as ineffective, and 4 percent are in unrated areas. Figure 7.1 presents a global overview of this coverage, and Figure 7.2 provides a summary for each region.

The very high levels of protection in Australia are clearly illustrated, with only one-quarter of reefs falling outside of MPA coverage. As a result, Australia skews the global averages considerably; outside of Australia MPA coverage is much lower, with only 17 percent of reefs inside MPAs. Of particular concern are the statistics for Southeast Asia, where only 3 percent of reefs are located within effective or partially effective MPAs.

FIGURE 7.2. REEF AREA BY MPA COVERAGE AND EFFECTIVENESS

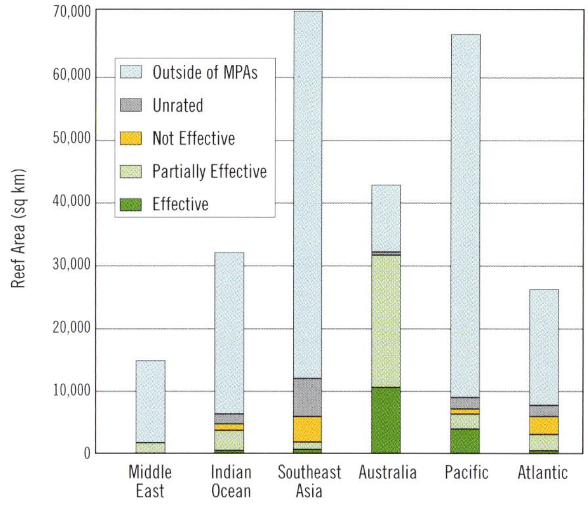

FIGURE 7.1. CORAL REEF-RELATED MARINE PROTECTED AREAS AND MANAGEMENT EFFECTIVENESS

A. Number of coral reef MPAs rated by management effectiveness

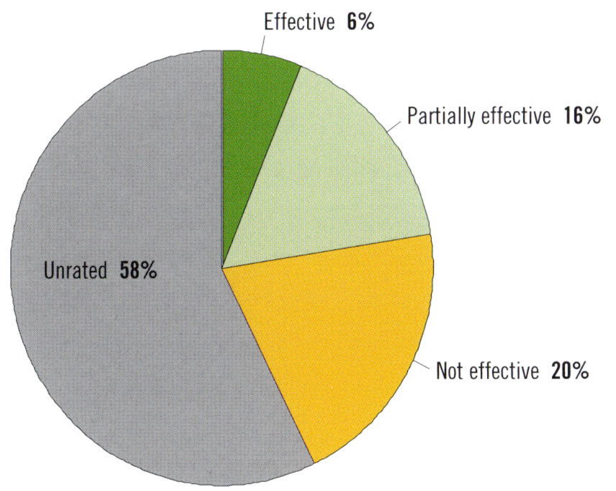

Note: The number of MPAs located in the coral reef regions of the world is approximately 2,679 (which represents 100% on this chart).

B. Coral reefs by MPA coverage and effectiveness level

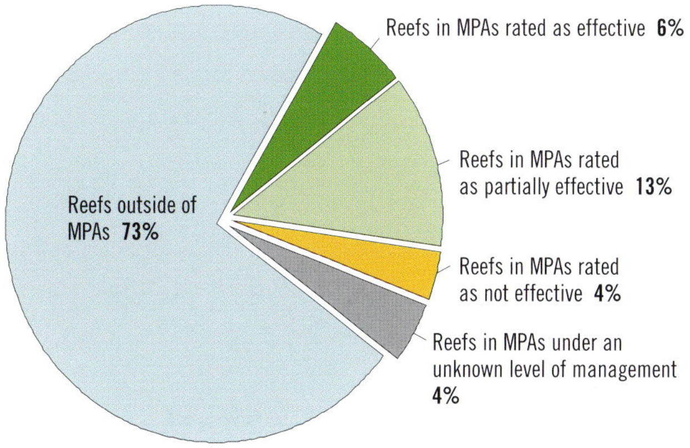

Note: The global area of coral reefs is 250,000 sq km (which represents 100% on this chart), of which 67,350 sq km (27%) is inside MPAs.

MPAs in the Reefs at Risk Model

MPAs were factored into the overfishing and destructive fishing components of the *Reefs at Risk Revisited* model, reducing the impact of unsustainable fishing for those areas where effective or partially effective management was in place. Although MPAs can and do serve many other functions for reefs beyond reducing fishing pressure (such as protection of mangroves or protection of land from development), these effects are less consistent, and thus were not included in the model. We found that MPAs did not have a large influence on the model of overfishing threat; MPAs, particularly large sites, are located disproportionately in areas of low fishing pressure, and management effectiveness tends to be lower in areas of high fishing pressure. As a result, MPAs only reduced the global level of overfishing by just over 2.5 percentage points. The Atlantic had the greatest reduction in overfishing due to MPAs, where the threat was reduced by over 4 percentage points.

The most dramatic influence of MPAs have been observed within no-take areas where all extractive activities are controlled.[61, 285] At the present time there is no complete data set describing which MPAs, or parts of MPAs, are no-take zones. However, an earlier (2008) review was able to assess more than 30 percent of the world's MPAs, covering (then) 65 percent of total area protected.[286,287] From this subset we are aware of 241 coral reef MPAs with total or partial no-take coverage, which includes about 12,150 sq km (4.9 percent) of the world's coral reefs. Such statistics are once again heavily influenced by Australia: outside of Australia only 1,920 sq km of reefs (less than 1 percent) are in no-take areas .[288]

BOX 7.2. MANAGING FOR CLIMATE CHANGE

PHOTO: FRED PAWA, TNC

One of the greatest challenges to coral reef conservation comes from climate change. Unlike other threats, damage to reefs from climate change cannot be prevented by any direct management intervention. However, there is good evidence that the likelihood and severity of damage on particular reef ecosystems can be reduced by 1) identifying and protecting areas of reef that are naturally likely to suffer less damage from climate change (that is, promoting "reef resistance"), and 2) designing management interventions to reduce local threats and improve reef condition, so that rates of recovery can be improved (that is, promoting "reef resilience").[119, 289] Reef resilience is the basis for a number of new tools designed to help managers deal with climate change.[118] It involves developing a management framework, centered on MPAs, but extending beyond them using some of the integrated approaches described earlier. Small, isolated MPAs are less likely to promote resilience than networks of MPAs, which would ideally include some large areas. MPA networks should include representation of all reef zones and habitats to reasonable extents. Furthermore, they must protect critical areas, such as fish spawning areas or bleaching-resistant areas. The networks should also be designed to utilize connectivity, so that replenishment following impacts can be maximized. Finally, it is critical to establish effective management to reduce or eliminate other threats that would otherwise hinder recovery.[290] Although the impacts of ocean acidification have still not been broadly shown *in situ*, it is possible that proposed measures for managing reefs in the context of warming seas may also provide better conditions for corals to survive early stages of ocean acidification. It is critical to note that, at best, such local-scale measures will only buy time for coral reefs—accelerating climate change will eventually and irreversibly affect all reef areas unless the ultimate cause of warming and ocean acidification, greenhouse gas emissions, is addressed by the global community.

Chapter 8. CONCLUSIONS AND RECOMMENDATIONS

PHOTO: STEVE LINDFIELD

This report portrays a deeply troubling picture of the world's coral reefs. Local human activities already threaten the majority of reefs in most regions, and the accelerating impacts of climate change are compounding these problems globally. Alongside our findings on reef pressures, the report also shows how coastal nations and communities around the world depend on precious reef resources for ecosystem services, notably food, livelihoods, and coastal protection, and highlights the significant economic values of these services.

These two aspects—high threat and high value—point to an obvious and urgent need for action. Here, the report offers some reason for hope: reefs around the world have shown a capacity to rebound from even very extreme damage, while active management is protecting reefs and aiding recovery in some areas. We have a clear understanding of the threats to coral reefs, and a growing appreciation of the additional complexities that arise from compounding threats. We have also identified and tested many of the approaches needed to effectively manage and safeguard reefs. Meanwhile, there is a strong desire from all sectors to secure a future for coral reefs and for the many communities and nations that rely upon them. What is needed now is action. Just as the pressures themselves are broad and interconnected, so the array of measures to deal with them must be comprehensive. Local threats, often pressing and immediate, must be tackled head-on with direct management interventions. Climate change poses a grave danger not only to reefs, but to nature and humanity as a whole; efforts to quickly and significantly reduce greenhouse gas emissions are of paramount concern. At the same time, we may be able to buy time for coral reefs in the face of climate change through local-scale measures to increase their resilience to climate-related threats.

Success for any management effort, from local to global, is highly dependent on obtaining the support of all parties involved. Fortunately, the actions to reduce threats to reefs are often "win-win" solutions that make both social and economic sense, in addition to providing broader environmental benefits.

Most of the threats discussed in this report are not new—the first *Reefs at Risk* report in 1998 also highlighted the risks posed by coastal development, watershed-based pollution, marine-based pollution, overfishing and destructive fishing, and coral bleaching due to high sea temperatures. That report also made recommendations to reverse the situation. Since 1998 there has been a remarkable increase in efforts to protect and sustainably manage coral reefs, exemplified by a fourfold increase in coral reef protected areas since 2000,[14] by improvements in management effectiveness and by increases in funding, scientific research, and community engagement. Clearly such efforts have not been enough, but neither is it correct to say we have failed. Threats have grown, both in intensity and extent, faster than the growth in efforts to manage these pressures, but coral reefs might have declined considerably more than they have without existing efforts.

We need to do much more, and do it faster. Our collective ability to do so has become stronger, with new management tools, increased public understanding, better communications, and more active local engagement. We hope that this new report will spur further action to save these critical ecosystems.

Below are recommendations for how to address each category of threats to reefs highlighted in this report; how to build consensus and capacity to drive change, and how to scale up existing efforts to match the challenge. We conclude with actions that individuals can take to help reverse the decline of these valuable and spectacular ecosystems. Links to useful resources on individual actions and management approaches to aid coral reefs can be found at www.wri.org/reefs.

MANAGE COASTAL DEVELOPMENT

The impacts associated with coastal development can be substantially reduced through effective, integrated coastal planning focused on sustainable development, coupled with enforcement of coastal development regulations.

- **Protect critical coastal habitats.** Dredging and landfilling activities should be avoided near reefs, where direct damage may occur. Mangroves and seagrasses, which are often in close proximity to coral reefs, also need to be protected from development. They trap sediments and nutrients, promoting clear waters near reefs, and serve as important nursery areas. Mangroves also protect shorelines from erosion and storm damage.

- **Establish and honor coastal development setbacks.** Restricting or limiting coastal development within a specified distance from the coast through "coastal development setbacks" is a sensible, precautionary approach that planning authorities can take to protect sensitive coastal habitat, and to lessen the risk of property loss through erosion or damage from storms. The growing climate-related threats of sea level rise and increased storm intensity provide further incentive.

- **Develop coasts with nature in mind.** Sensible planning can reduce the potential for extensive damage to reefs during and after construction. Land developers should use sediment traps (physical barriers that prevent eroded soil from entering coastal waters) during construction and avoid development in steep areas, to lessen erosion impacts. Landfilling (land reclamation) in shallow waters near reefs, including all reef flats, should be prevented. In addition, construction materials should not be obtained by mining corals or sand near living reefs. Coastal development practices should also include planning and infrastructure to control storm water, to avoid drainage and erosion problems after construction is complete.

- **Manage wastewater.** To maintain coastal water quality and reduce the nutrients and toxins that reach coral reefs, wastewater (including sewage and industrial effluent) must be treated and controlled. Ideally, sewage should be treated to the tertiary level (that is, a high level of nutrient removal); however, such treatment is often too costly for many coastal communities without the help of outside donors. A less expensive interim solution is simply to manage the flow and release of wastewater. Such management options include directing effluent to settling ponds for natural filtering by vegetation, or routing discharge far offshore, well beyond reefs.

- **Link terrestrial and marine protected areas.** Marine protected areas (MPAs) can be highly effective in reducing direct impacts such as overfishing, and where such sites are linked to terrestrial protected areas, they can have a considerable influence on reducing coastal development and watershed-based threats such as pollution and sedimentation. In addition, they can protect critical adjacent systems such as mangroves, coastal vegetation and lowland forests.

Mangroves are vital nursery areas for fish, and filter water and sediments coming off the land. Their maintenance or restoration can be a critical component of coral reef conservation.

- **Implement tourism sustainably.** Poorly planned tourism can severely damage coral reefs and undermine the very environments that attract visitors. Implementing sustainable tourism practices, such as those outlined below, can promote long-term benefits for reefs, the local economy, communities, and the tourism industry.
    - *Follow the rules.* As with other coastal development, it is important to honor coastal development regulations when building for tourism. It is critical to retain coastal habitats (including beach shrubs, mangroves, and seagrasses), source construction materials sustainably, honor coastal setbacks, and provide infrastructure to treat sewage and waste.
    - *Manage marine recreation.* Installing and using mooring buoys can significantly reduce anchor damage to reefs from boats. In addition, dive operators should evaluate and respect the carrying capacity of dive sites, which will both lessen the impact from recreational use and improve the tourist experience.
    - *Source sustainably.* Tourists create additional demand for seafood and souvenirs. Businesses should obtain such products from environmentally and socially responsible suppliers, and shops and restaurants should avoid selling corals or serving seafood that is not sustainably harvested.
    - *Engage communities.* Tourism that involves and benefits local communities is more sustainable in the long run. Hotels, restaurants, and tour operators should make every effort to ensure benefits are felt locally, such as through employment practices, supporting local industry, and building positive connections between visitors and local people.
    - *Engage the tourism industry.* Partnerships between the private sector and governments or NGOs can facilitate information exchange, training in best environmental practices, and collaborative efforts to find solutions to issues of shared concern. Such partnerships can also be economically beneficial for tourist providers, by increasing their attractiveness to tourists and operators who prefer environmentally responsible options.
    - *Encourage good practices.* Certification schemes, accreditation, and awards facilitate best practices for hotels, dive operators, and tour operators. These incentives encourage eco-friendly development, while also attracting high-end tourists.
    - *Go beyond the rules.* Recognizing that rules may be lacking or insufficient to ensure a secure future for coral reefs, tourist facilities should be proud leaders of environmental protection and should set standards above and beyond those required in legal or advisory texts.
    - *Educate tourists.* Raising tourist awareness about the local importance of coral reefs will increase the likelihood that they will treat the ecosystem respectfully during their visit, as well as advocate for reef conservation in their home countries.

Integrated watershed management, including preserving and restoring forests on steep slopes and river margins, helps to reduce sediment and nutrients reaching coral reefs.

REDUCE WATERSHED-DERIVED SEDIMENTATION AND POLLUTION

Integrated watershed management can help to lessen land-derived pollution and sediment. Other key sectors, such as agriculture and forestry, need to be actively engaged in such efforts if realistic and sustainable solutions are to be achieved.

- **Manage watersheds to minimize nutrient and sediment delivery.** Avoiding cultivation of steep slopes, retaining or restoring riparian vegetation, and managing soil to minimize loss and nutrient runoff are essential. Conservation tillage, soil testing to determine appropriate fertilizer need, reduction or elimination of subsidies that promote excessive fertilizer application, and appropriate application of pesticides can all help to mitigate pollution of coastal waters and damage to coral reefs.

- **Manage livestock waste.** To reduce nutrient and bacterial pollution of coastal waters, farmers should keep livestock away from streams and use settling ponds to manage animal waste.

- **Control grazing intensity.** At high densities, particularly on steeper slopes, livestock can greatly reduce vegetation cover and increase erosion, adding sediment loads to coastal waters. This is a problem even on remote islands where feral ungulates (sheep, goats, and pigs) are uncontrolled. Lower stocking densities, avoiding grazing on steeper slopes and the active control of feral animals help relieve this threat.

- **Retain and restore vegetation.** Terrestrial protected areas can be used as a tool to protect sensitive habitats within watersheds, including riparian vegetation and steep slopes. Preserving and restoring natural vegetation, such as forests, can help reduce erosion and return areas to a stable state.

- **Control runoff from mines.** Mining for materials such as sand, gravel, metals, or minerals near the coast can be a significant source of sediment, rubble, heavy metals, and other toxic pollutants in nearshore waters. Controlling erosion and runoff around mines using stormwater diversion channels, sediment traps, and settling ponds can reduce mining impacts on coastal water quality.

REDUCE MARINE-BASED SOURCES OF POLLUTION AND DAMAGE

Pollution and damage from ships are constant threats to coral reefs, especially in high-traffic areas like ports and marinas. Likewise, oil and gas development continues to expand in both coastal and offshore waters.

- **Improve waste management at ports and marinas.** By installing or expanding waste disposal and treatment facilities, ports and marinas can significantly reduce the amount of trash, wastewater, and bilge water that vessels send overboard. Wastewater should be treated to the tertiary level.

- **Control ballast discharge.** Governments should establish policies that limit exchange of ballast water to the deep ocean, in order to minimize the uptake and establishment of invasive species in sensitive shallow waters. Ports and marinas should also install ballast water reception and treatment facilities to eliminate ballast discharge at port.

- **Designate safe shipping lanes and boating areas.** National governments should restrict the areas where ships are permitted to navigate to deep waters, to protect reefs from large-vessel groundings. Establishing areas where vessels of all sizes may safely drop anchor and restricting all vessels from entering certain critical habitats will also protect reefs from physical damage.

- **Manage offshore oil and gas activities.** Governments and the private sector should take precautions against accidents and spillages from coastal and offshore oil and gas activities, establish policies that mitigate such risks, and prepare adequate emergency response plans that will minimize impact. Governments should also consider designating critical habitats as off-limits to drilling, dredging, and other related oil and gas exploration activities.
- **Use MPAs to protect reefs and adjacent waters.** MPA regulations need to include restrictions on shipping traffic, and controls on physical damage and vessel discharge. The expansion of MPAs in broad buffers around reefs can provide further protection and secure important related ecosystems such as seagrass beds and mangrove forests.

REDUCE UNSUSTAINABLE FISHING

The impacts of overfishing and destructive fishing—the most widespread of all threats to reefs—can be reduced through proper management of fishing areas and practices, alongside efforts to recognize and redress underlying social and economic factors.

- **Address the underlying drivers of overfishing and destructive fishing.** Food insecurity, poverty, poor governance, conflicts with other resource users, and a lack of alternative livelihoods within a community or nation can all contribute to overfishing. Identifying and addressing these drivers, and helping to establish sustainable alternative or additional sources of dietary protein and income (e.g., eco-tourism or aquaculture) can alleviate pressure on fisheries, and increase the likelihood that other management measures will succeed.
- **Manage fisheries.** Government fisheries agencies should work with resource users to establish sustainable policies and practices. These measures may involve regulating fishing locations, seasons, gear types, or catch numbers, which can reduce fishing pressure on reef species. Specific approaches include rotating fishing ground closures, establishing a finite number of fishing licenses, banning damaging fishing gears (such as trawl and seine nets that destroy benthic habitats), establishing minimum size restrictions, or limiting catch numbers to a sustainable level.
- **Reduce excessive fishing capacity and remove unsound fishing subsidies.** Perverse incentives such as subsidizing boat purchases, fuel, fishing gear, or catch values encourage unsustainable fishing and distort market forces that might otherwise regulate the size of the fishing industry. Governments should eliminate such incentives, and redirect investment toward developing alternative livelihoods and reducing overcapacity.
- **Halt destructive fishing.** Fishing with explosives and poisons is illegal in many countries, but persists due to a lack of enforcement. Increasing the capacity of fisheries management authorities to enforce laws that ban fishing with explosives and poisons, as well as establishing strict penalties, will reduce fishers' incentives to use these illegal methods. Other strategies include educating fishers and other local stakeholders about the long-term consequences of destroying critical habitat and the personal risks of using these hazardous methods.
- **Expand MPAs to maximize benefits.** Increasing the area of coral reefs that are located inside MPAs (especially inside designated no-take zones that prohibit fishing) helps protect fish habitats and replenish depleted stocks. Expansion of MPAs should reflect a regional perspective, recognizing the interdependence of reef communities and the transboundary nature of many reef threats. When locating new MPAs, planners should consider biodiversity, resilience to disturbances, connectivity, and other characteristics that may maximize the benefits of protection. The development of MPA networks, or of very large-zoned MPAs, utilizing ecological and socioeconomic knowledge can create systems that are considerably more effective in supporting productivity and resisting or recovering from stress, than sites declared without such planning. MPAs are currently underutilized in areas where human pressures are greatest. A key priority for governments should be to accelerate MPA or equivalent

The complete closure of even small areas to fishing (no-take areas) can lead to rapid reef recovery, as well as to improved fishing in surrounding areas.

local management designations in such places, in collaboration with local stakeholders.

- **Improve the effectiveness of existing MPAs.** MPAs require day-to-day management and enforcement to effectively protect reef resources, yet many exist only on paper and lack the economic resources and staff for effective management. Governments, donors, NGOs, and the private sector should provide financial and political support to help MPAs build needed capacity, both in terms of equipment (e.g., boats and fuel) and adequately trained staff. MPAs are more likely to be successful in the long term if they are financially self-sustaining, with a diverse revenue structure, and many will require further support to achieve this aim.
- **Involve stakeholders in resource management.** Community inclusion and participation in decisions that affect reef resources are critical to establishing the acceptance and longevity of management policies. When "top-down" decisions are made without community consultation, local knowledge and capacity are left untapped, and programs may fail to respond to the needs of users or to win their support. Governments, resource managers, and NGOs should promote community and other stakeholder involvement in both decisionmaking and management of resources.

MANAGE FOR CLIMATE CHANGE AT LOCAL SCALES

At local scales, it is vital to maintain and promote reef resilience, to encourage faster recovery after coral bleaching, and increase the likelihood that coral reefs will survive climate change. Such efforts may represent an opportunity to buy time for reefs, until such time as global greenhouse gas emissions can be curbed (see *Scaling up: International Collaboration,* below).

- **Build resilience into planning and management.** Building resilience to climate change requires reducing local threats, including overfishing, nutrient and sediment pollution, and direct physical impacts to reefs. As not all sites can be protected, it makes sense to target critical areas, such as fish spawning areas or sites that supply other reefs with coral larvae, which will support replenishment of other areas. Because the location of individual future stress events cannot be predicted, it is also critical to ensure representative areas are included in protected areas systems, and that there is replicate protection at multiple sites.
- **Explore options for local-scale reef-focused interventions.** In the event that conditions become untenable for reefs, a wide range of potential management measures have been suggested. These approaches include *ex-situ* conservation of reef species in aquaria, artificial shading to reduce temperature stress, deploying powdered lime to locally reduce acidity, and genetic manipulation of zooxanthellae (the microscopic algae within corals) or other sensitive species. The unforeseen consequences of any such radical interventions call for extreme caution, but it may be wise to begin the debate about these and other interventions before the pressure for action becomes more intense.

ACHIEVING SUCCESS

The broad array of strategies and interventions outlined above point to the need for one critical further action:

- **Develop cross-sectoral, ecosystem-based management efforts.** To avoid duplicated or conflicting management approaches, and to maximize the potential benefits from efforts to protect the coastal zone, it is critical to develop plans that are agreed upon by all sectors and stakeholder groups, and that take into account ecological reality. Ecosystem-based management is a term used to encourage the understanding and utilization of natural patterns and processes into overall planning. Integrated coastal management (ICM), ocean zoning, and watershed management are all terms that are widely used and being increasingly applied to encourage such connected thinking.

BUILDING CONSENSUS AND CAPACITY

Knowledge about reef species, threats, and management approaches has grown tremendously in recent years, allowing reef users and managers to better recognize problems, address threats, and gain political, financial, and public support for reef conservation. Nevertheless, a gap remains between our existing knowledge and results. Closing this gap depends on action within the following key areas:

- **Research** to further develop the body of evidence showing how particular reefs are affected by local activities and climate change and how different stressors may act in combination to affect reef species; identifying factors that confer resilience to reef species and systems; determining the extent of human dependence on specific reef ecosystem services; and understanding the potential for reef-dependent nations and communities to adapt to expected change.
- **Education** to inform communities, government agencies, donors, and the general public about how current activities threaten reefs and those who rely on them and why preventive action is needed.
- **Communication** to spread the message that action is urgently needed to save reefs and to highlight examples of conservation success that can be replicated (See *Scaling Up*).

Education and communication are essential to identifying appropriate solutions and building stakeholder support.

- **Economic valuation** to highlight the worth of reefs, as well as the scale of economic and social losses that will result if reefs degrade. Valuation also provides a tool for evaluating the costs and benefits of management and development options, with an emphasis on long-term benefits, which can help avoid short-sighted development.
- **Policy support** to aid decisionmakers and planners in making long-term decisions that affect the survival of coral reefs and the ability of coastal people to adapt to changes associated with coral degradation and rising seas.
- **Training and building capacity** of reef stakeholders, to manage and protect reef resources. Building capacity for reef management and law enforcement among local communities, agencies and organizations can directly benefit reef resources. Developing familiarity with economic concepts can help stakeholders to understand and argue for the important benefits that reefs provide. Training in communication and education will help to spread the awareness needed to change human behavior. In addition, supporting and training fishers in alternative or additional livelihood activities, where appropriate, can help take pressure off of reefs and reduce vulnerability in reef-dependent regions.
- **Involvement** of local stakeholders in the decisionmaking and management of reef resources is critical to the development of successful plans and policies.

SCALING UP: INTERNATIONAL COLLABORATION

We already have much of the knowledge, information, and tools needed to take actions that will effectively reduce local pressures on coral reefs and promote reef resilience in the face of a changing climate. However, at both local and national scales, we also need the political will and economic commitment to implement these actions. If we are to achieve meaningful results globally, it is critical to scale up these local and national approaches, and work internationally: to share knowledge, experience and ideas; to seek solutions to global-scale threats; and to make use of existing international frameworks to foster change. Examples of such international tools include:

- **International agreements.** When signed, ratified, and enforced, international agreements are important tools for setting and achieving collective goals. Agreements in several key areas may help to reduce threats to reefs. The UN Convention on the Law of the Sea establishes ocean governance which, when used effectively and enforced, can significantly reduce fishing pressure in domestic waters. CITES is an effective international agreement designed to control the trade of listed endangered species, including most hard coral species. MARPOL provides a framework for minimizing marine pollution from ships, but more widespread adoption and enforcement is needed in coral reef nations. The UN Framework Convention on Climate Change (UNFCCC) provides an important framework and urgently needs to establish new strict and binding protocols to drive reductions in greenhouse gas emissions. Climate adaptation funds, such as those established under the UNFCCC, should support efforts to protect coral reefs and reduce vulnerability of reef-dependent people (e.g., through livelihood enhancement and diversification), as key priorities for adaptation planning.

- **Transboundary collaboration and regional agreements.** Neither marine species nor pollution respect political boundaries. Efforts to effectively manage coral reefs and reduce pressures will often be transboundary in nature—including managing reef fisheries and trade, establishing international MPAs, and managing river basins that cross political borders. Regional agreements such as the Cartagena Convention (to address land-based sources of pollution, oil spills, protected areas, and wildlife) may have a pivotal role to play in achieving political commitments. Elsewhere, smaller bilateral or multilateral agreements may suffice, building trust and enabling the sharing of experience, resources, and results to build effective management up to larger scales.

- **International regulations on the trade in reef products.** Better regulation is needed on all trade in reef products. In particular, trade in live reef organisms should require certification to show that they have been sustainably caught, using nondestructive methods, and that they have been held and transported in a way that minimizes mortalities. In order to achieve this goal, testing and monitoring must be improved at the national level, to reliably identify sustainably fished or aquacultured species from those that have been harvested unsustainably or illegally. Cyanide detection facilities should be established at major live fish collection and transshipment points, for both the live reef food fish trade and the aquarium trade. Monitoring should include assessments of shipping practices and holding facilities along the supply chain.

- **Climate change efforts.** Coral reefs are extremely sensitive to climate change, which has led many reef scientists to recommend not only a stabilization of CO_2 and other greenhouse gas concentrations, but also a longer-term reduction to 350 ppm.[94] This target will be extremely challenging to attain, requiring immense global efforts to reduce emissions and, possibly, to actively remove CO_2 from the atmosphere. These actions will only be driven by demand, by reason, and by example. Thus there is a role to be played by all—individuals and civil society, NGOs, scientists, engineers, economists, businesses, national governments, and the international community—to address this enormous and unprecedented global threat.

INDIVIDUAL ACTION: WHAT YOU CAN DO

Many of the recommendations outlined in this chapter involve collaboration among multiple sectors and require political will, which can take time and coordinated efforts to achieve. However, immediate results are also attainable from individual actions.

If you live near a coral reef:

- **Follow the rules.** Learn about local laws and regulations designed to protect reefs and marine species, and obey them. Setting a good example can encourage a broader sense of environmental stewardship in your community.

- **Fish sustainably.** If you fish, for food or recreation, try to minimize your impact. Never take rare species, juveniles, or undersized species, or those that are breeding or bearing eggs. Do not fish in spawning aggregations and do not take more than you need. Never abandon fishing gear.

- **Avoid physical damage.** Boat anchors, trampling, and even handling corals can damage the structure of the reef environment. Try not to touch.

- **Minimize your indirect impacts on reefs.** Pay attention to where your seafood comes from, and how it is fished; know which fish are caught sustainably and which are best avoided. Find out whether your household waste and sewage are properly disposed of, away from the marine environment, and if not, press for change. Reduce your use of chemicals and fertilizers, to prevent these pollutants from entering inshore waters.

- **Help improve reef protection.** If there are insufficient conservation measures around your area, work with others to establish more. Be aware of planned development projects in coastal or watershed areas, and participate in public consultation processes. Support local organizations and community groups that take care of the reefs. Organize or help with reef, beach, and watershed clean-ups, reef monitoring, restoration, and public awareness activities.

Getting involved in local activities, such as replanting mangroves, can be enjoyable and can also benefit the local environment.

- **Campaign for the reefs.** Let your local political representative or government know that the reefs are important to you and your community. Political will is important for establishing policies to better manage and protect reefs, to support the people and businesses that rely upon them, and to address the serious threats posed by climate change. Creating such policies begins with making reef issues a priority among political decisionmakers.

If you visit coral reefs:

- **Ask before you go.** Find out which hotels and tourism operators at your destination are sustainably managed and eco-conscious (that is, treat their wastewater; support local communities; honor coastal setbacks) and patronize them.

- **Dive and snorkel carefully.** Touching corals with your hands, body, or fins can damage delicate reef structures, potentially harming both you and the reef. Keep a short distance away. As the saying goes: take only pictures and leave only bubbles.

- **Tell people if they are doing something wrong.** If you see someone littering on a beach or in the sea, or stepping on corals… say something! Showing that you care and informing others can propagate good environmental stewardship.

- **Visit MPAs and make a contribution.** Many MPAs acquire funding for management through fees and donations from tourists. If you are vacationing near an MPA, support its reef management by visiting, diving, or snorkeling there.

- **Avoid buying coral souvenirs.** Purchasing products made from corals or other marine organisms encourages excessive (and in some cases, illegal) harvesting of such resources. As an informed tourist, refrain from purchasing souvenirs made from marine species.

Wherever you are:

- **Eat sustainably.** Choose to eat sustainably caught seafood and avoid overfished species like groupers, snappers, and sharks. Many seafood products that are sustainably sourced bear an eco-certification label, from organizations including the Marine Stewardship Council (www.msc.org) and Friend of the Sea (www.friendofthesea.org). Wallet-sized guides and mobile phone applications for making informed seafood choices are available from organizations such as the Australian Marine Conservation Society (www.marineconservation.org.au), Monterey Bay Aquarium's Seafood Watch (www.seafoodwatch.org), Sea Choice (www.seachoice.org), Southern African Sustainable Seafood Initiative (www.wwfsassi.co.za), and WWF (www.panda.org).

- **Avoid buying marine species that are threatened, or that may have been caught or farmed unsustainably.** Purchasing wildlife that is already under threat encourages continued overexploitation, putting these species at greater risk. Be an educated consumer—purchase items from reputable vendors, and ensure that animals purchased live for aquariums are certified, such as through the Marine Aquarium Council (MAC) (www.aquariumcouncil.org).

- **Vote for conservation.** Be an informed voter and know the priorities of your government representatives. If the environment, conservation, and climate change are not major issues for them, call or write to them to voice your opinion and make these issues a priority.

- **Support NGOs that conserve reefs and encourage sustainable development in reef regions.** Many NGOs around the world support reef conservation on local, regional, or global scales. Contributing time or money to these organizations can help them in their efforts to conserve reef resources, encourage sustainable development in reef regions, and support reef-dependent communities.

- **Educate through example.** People are most influenced by their friends, family, and peers. Showing people that you care about reefs, and helping them to appreciate why reefs are important to you—by sharing books, articles, or videos—are important steps in propagating a conservation message.

- **Reduce your CO_2 emissions.** Changes to climate and to ocean chemistry may soon become the greatest threats of all to coral reefs. Individual actions will never be enough, but reducing our individual carbon footprints sends a powerful message to those we know, and to those we vote for.

Whichever of these you do, encourage others to do the same.

CONCLUSION

Coral reefs are vital to coastal communities and nations around the world. They are a source of inspiration to many more. The threats to the world's coral reefs, however, are serious and growing. This report has portrayed the precarious state of coral reefs globally, encroached upon from all sides by numerous threats. In the face of such pressures it is critical that we focus on practical, immediate responses, such as those highlighted above, to reduce and to reverse these threats. We are at a critical juncture. We know what is needed. Action now could ensure that coral reefs remain, and that they continue to provide food, livelihoods, and inspiration to hundreds of millions of people now, and for generations into the future.

Appendix 1. MAP OF REEF STORIES

MAP A.1. LOCATIONS OF REEF STORIES

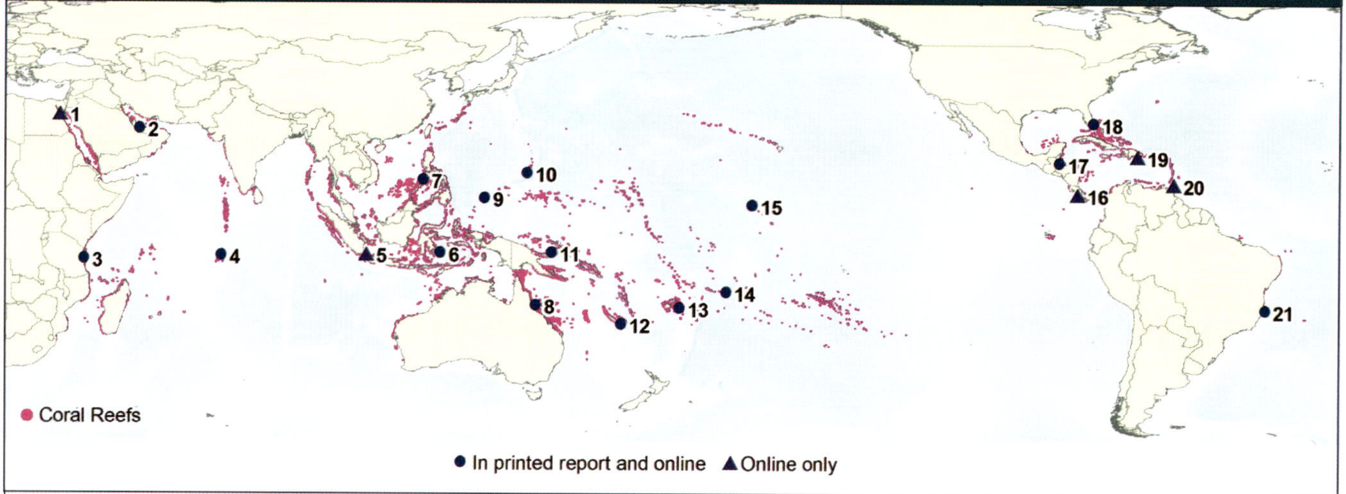

Notes: Map presents the locations of Reef Stories—case studies of particular coral reefs, their threats, and management—that were submitted by reef managers around the world. As noted in the map legend, short versions of many stories are in this report, and all are located on the Reefs at Risk website along with longer versions that include additional details. See www.wri.org/reefs/stories. The numbers in the map correspond to the table below.

Number on Map	Reef Story	Located on:
1	Egypt: Coral Survival Under Extreme Conditions	Website
2	Persian Gulf: The Cost of Coastal Development to Reefs (Box 5.1)	Page 50
3	Tanzania: Deadly Dynamite Fishing Resurfaces (Box 3.5)	Page 26
4	Chagos Archipelago: A Case Study in Rapid Reef Recovery (Box 5.2)	Page 53
5	Indonesia: New Hope for Seribu Island's Reefs	Website
6	Indonesia: People Protect Livelihoods and Reefs in Wakatobi National Park (Box 5.3)	Page 55
7	Philippines: Social Programs Reduce Pressure on Culion Island's Reefs (Box 6.2)	Page 75
8	Australia: Remaining Risks to the Great Barrier Reef (Box 5.4)	Page 58
9	Palau: Communities Manage Watersheds and Protect Reefs (Box 3.3)	Page 24
10	Guam: Military Development Threatens Reefs (Box 3.1)	Page 22
11	Papua New Guinea: Marine Protection Designed for Reef Resilience in Kimbe Bay (Box 3.9)	Page 33
12	New Caledonia: Reef Transplantation Mitigates Habitat Loss in Prony Bay (Box 5.6)	Page 62
13	Fiji: Local Management Yields Multiple Benefits at the Namena Marine Reserve (Box 7.1)	Page 81
14	American Samoa: Shipwreck at Rose Atoll National Wildlife Refuge (Box 3.4)	Page 25
15	Line Islands: A Gradient of Human Impact on Reefs (Box 5.5)	Page 61
16	Costa Rica: Reef Life after Bleaching	Website
17	Mesoamerican Reef: Low Stress Leads to Resilience (Box 3.7)	Page 30
18	Florida: Marine Management Reduces Boat Groundings (Box 5.7)	Page 65
19	Dominican Republic: Protecting Biodiversity, Securing Livelihoods at La Caleta National Marine Park	Website
20	Tobago: A Sustainable Future for Buccoo Reef	Website
21	Brazil: Coral Diseases Endanger Reefs (Box 3.10)	Page 37

Appendix 2. DATA SOURCES USED IN THE *REEFS AT RISK REVISITED* ANALYSIS

Data used in the *Reefs at Risk Revisited* threat analysis, model results, metadata, and full technical notes on the modeling method are available on CD by request and online at www.wri.org/reefs. Data used in the analysis are outlined below, by category of threat:

COASTAL DEVELOPMENT

- **Population density**—LandScan (2007)™ High Resolution global Population Data Set copyrighted by UT-Battelle, LLC, operator of Oak Ridge National Laboratory under Contract No. DE-AC05-00OR22725 with the United States Department of Energy.
- **Population growth**—Derived at WRI from LandScan (2007) and Global Rural-Urban Mapping Project (GRUMP), Alpha and Beta Versions: Population Density Grids for 2000 and 2005. GRUMP is a product of the Center for International Earth Science Information Network (CIESIN), Columbia University; International Food Policy Research Institute (IFPRI); The World Bank; and Centro Internacional de Agricultura Tropical (CIAT). Palisades, NY: Socioeconomic Data and Applications Center (SEDAC), Columbia University.
- **Tourism data (tourist arrivals in millions)**—Development Data Group, The World Bank. World Development Indicators 2000 to 2006. Washington, DC: The World Bank, 2008.
- **City size and location**—Gridded Rural-Urban Mapping Project (GRUMP), 2005 (see above).
- **Ports**—National Geospatial Intelligence Agency, World Port Index, 2005.
- **Airports**—Digital Aeronautical Fight Information File (DAFIF), a product of the National Imagery and Mapping Agency (NIMA) of the United States Department of Defense (DOD), 2006.
- **Hotels and resorts**—Provided by HotelbyMaps www.hotelbymaps.com, 2009, and downloaded for select countries from GeoNames www.geonames.org, 2010.

Special thanks to Daniel Hesselink and Qyan Tabek (HotelbyMaps.com), Gregory Yetman (CIESIN), Azucena Pernia (WTO), and Siobhan Murray and Uwe Deichmann (World Bank) for their assistance in compiling these data sets.

WATERSHED-BASED POLLUTION

- **Watershed boundaries**—Based on HydroSHEDS (15 arc-second/500 meter resolution) produced by the World Wildlife Fund in partnership with the U.S. Geological Survey (USGS), the International Centre for Tropical Agriculture (CIAT), The Nature Conservancy (TNC), and the Center for Environmental Systems Research (CESR) of the University of Kassel, Germany. Available at: http://hydrosheds.cr.usgs.gov. For Pacific Islands: derived at WRI from NASA/NGA, Shuttle Radar Topography Mission (SRTM) Digital Elevation Model (DEM) (3 arc-second/90 meter resolution).
- **Landcover data**—ESA/ESA GlobCover Project, led by MEDIAS-France, 2008 coupled with agricultural areas from Global Land Cover Database (GLC2000), EU Joint Research Centre 2003.
- **Precipitation**—Data are from Berkeley/CIAT/Rainforest CRC, www.WorldClim.org, Average Monthly Precipitation 1950–2000, version 1.4, 2006.
- **Soil porosity**—FAO/IIASA/ISRIC/ISS-CAS/JRC. Harmonized World Soil Database (version 1.0). FAO, Rome, Italy and IIASA, Laxenburg, Austria, 2008.
- **Dams**—Global Water System Project. Global Reservoir and Dam (GRanD) Database, 2008.
- **Great Barrier Reef plumes**—Devlin, M., P. Harkness, L. McKinna, and J. Waterhouse. 2010. *Mapping of Risk and Exposure of Great Barrier Reef Ecosystems to Anthropogenic Water Quality: A Review and Synthesis of Current Status.* Report to the Great Barrier Reef Marine Park Authority. Townsville, Australia: Australian Centre for Tropical Freshwater Research.

Special thanks to Ben Halpern (NCEAS), Shaun Walbridge (UCSB), Michelle Devlin (JCU), Carmen Revenga (TNC), and Bart Wickel (WWF) for their assistance in compiling these data and advising on modeling.

MARINE-BASED POLLUTION AND DAMAGE

- **Port volume, commercial shipping activity, and oil infrastructure** data provided by Halpern, et al. 2008. "A Global Map of Human Impact on Marine Ecosystems." *Science* 319: 948-952. Original sources are as follows:
  - **Ports**—National Geospatial Intelligence Agency, World Port Index, 2005.
  - **Commercial shipping lane**s—World Meteorological Organization Voluntary Observing Ships Scheme, 2004–05.
  - **Oil infrastructure**—Stable Lights of the World data set prepared by the Defense Meteorological Satellite Program and National Geophysical Data Center (NGDC) within the National Oceanic and Atmospheric Administration (NOAA), 2003.
- **Cruise port and visitation intensity data** provided by Clean Cruising www.cleancruising.com.au based on compiled booking and departure information for all major cruiselines for July 2009 to June 2010.

Special thanks to Dan Russell (Clean Cruising), Ben Halpern (NCEAS), and Florence Landsberg (WRI) for their assistance in compiling these data sets.

OVERFISHING AND DESTRUCTIVE FISHING

- **Population density**—LandScan (2007)™ High Resolution global Population Data Set copyrighted by UT-Battelle, LLC, operator of Oak Ridge National Laboratory under Contract No. DE-AC05-00OR22725 with the United States Department of Energy.
- **Shelf area**—Derived at WRI from GEBCO Digital Atlas published by the British Oceanographic Data Centre on behalf of IOC and IHO, 2003.
- **Coral reef area**—IMaRS/USF, IRD, UNEP-WCMC, The World Fish Center and WRI, 2011. Global Coral Reefs composite data set compiled from multiple sources, incorporating products from the Millennium Coral Reef Mapping Project prepared by the Institute for Marine Remote Sensing, University of South Florida (IMaRS/USF), and Institut de Recherche pour le Développement (IRD/UR 128, Centre de Nouméa).
- **Population centers/market centers**—Gridded Rural Urban Mapping Project (GRUMP), Center for International Earth Science Information Network (CIESIN), Columbia University; and Centro Internacional de Agricultura Tropical (CIAT), 2005.
- **Destructive fishing**—Compiled at WRI from Reef Check surveys (2009), Tanzanian Dynamite Fishing Monitoring Network (2009), and expert opinion.

Special thanks to Gregor Hodgson and Jenny Mihaly (Reef Check); Elodie Lagouy (Reef Check Polynesia); Christy Semmens (REEF); Hugh Govan (LMMA Network); Annick Cros, Alan White, Arief Darmawan, Eleanor Carter, and Andreas Muljadi (TNC); Ken Kassem, Sikula Magupin, Cathy Plume, Helen Fox, and Lida Pet-Soede (WWF); Melita Samoilys (CORDIO); Ficar Mochtar (Destructive Fishing Watch Indonesia); Daniel Ponce-Taylor and Monique Mancilla (Global Vision International); Patrick Mesia (Solomon Islands Dept. of Fisheries); and the Tanzania Dynamite Fishing Monitoring Network, especially Sibylle Riedmiller (coordinator), Jason Rubens, Lindsey West, Matt Richmond, Farhat Jah, Charles Dobie, Brian Stanley-Jackson, Isobel Pring, and John Van der Loon.

GLOBAL THREATS: OCEAN WARMING AND ACIDIFICATION

Past Thermal Stress

- **Bleaching observations between 1998 and 2007**—ReefBase and UNEP-WCMC, WorldFish Center, 2009.
- **Thermal stress between 1998 and 2007**—National Oceanic and Atmospheric Administration, Coral Reef Watch, Degree Heating Weeks data (calculated from NOAA's National Oceanographic Data Center Pathfinder Version 5.0 SST dataset), http://coralreefwatch.noaa.gov, 2010.

Future Thermal Stress

- **Future thermal stress for decades 2030 and 2050**—Adapted from Donner, S. 2009. "Coping with Commitment: Projected Thermal Stress on Coral Reefs under Different Future Scenarios." *PLoS ONE* 4: e5712.

Ocean Acidification

- **Aragonite saturation state for the present, 2030, and 2050**—Adapted from Cao, L. and K. Caldeira. 2008. "Atmospheric CO_2 Stabilization and Ocean Acidification." *Geophysical Research Letters* 35: L19609.

Special thanks to Moi Khim Tan (WorldFish), Long Cao and Ken Caldeira (Stanford), Simon Donner (UBC), Ellycia Harrould-Kolieb (Oceana), Joan Kleypas (NCAR), Tim McClanahan (WCS), Joseph Maina (WCS), David Obura (IUCN), Elizabeth Selig (CI), and Tyler Christensen, Mark Eakin, Kenneth Casey, Scott Heron, Dwight Gledhill, and Tess Brandon (NOAA) for their assistance in providing thermal stress and acidification data and information.

CORAL REEF LOCATIONS

- **Coral reef map**—Institute for Marine Remote Sensing, University of South Florida (IMaRS/USF), Institut de Recherche pour le Développement (IRD), UNEP-WCMC, The World Fish Center and WRI, 2011. Global coral reefs composite data set compiled from multiple sources, incorporating products from the Millennium Coral Reef Mapping Project prepared by IMaRS/USF and IRD.

The coral reef location data were compiled from multiple sources by UNEP-WCMC, the WorldFish Center, and WRI. To standardize these data for the purposes of the *Reefs at Risk Revisited* project, data were converted to raster format (ESRI grid) at 500-m resolution. The original sources for the data include:

- Institute for Marine Remote Sensing, University of South Florida (IMaRS/USF) and Institut de Recherche pour le Développement (IRD). "Millennium Coral Reef Mapping Project," 2009 (30 m Landsat data classified and converted to shapefile).
- UNEP-WCMC. "Coral Reef Map," 2002.
- Additional data have been acquired or digitized from a variety of sources. Scales typically range from 1:60,000 to 1:1,000,000. See the technical notes and metadata online at www.wri.org/reefs for the full range of data contributors.

Special thanks to Corinna Ravilious (UNEP-WCMC), Moi Khim Tan (WorldFish), Serge Andrefouet (IRD), Julie Robinson (NASA), Frank Muller-Karger (USF), Stacy Jupiter and Ingrid Pifeleti-Qauqau (WCS-Fiji), and Jamie Monty (Florida DEP) for their assistance in compiling these data.

THREAT MODEL CALIBRATION

The modeling method for each threat component was developed in collaboration with project partners who provided input on threat indicators and preliminary thresholds. Each threat component was then calibrated individually, using available data from surveys and through review by project partners. The calibration guided the selection of thresholds between threat classes that are applied on a global basis. As such, Reefs at Risk indicators (for individual threats and the integrated local threat index) remain globally consistent indicators.

A range of monitoring and assessment data were used to explore patterns of coral reef degradation and calibrate the threat analysis:

- **Reef Check**—Volunteer survey program that has collected biophysical data at reef sites for more than 3,000 survey sites in about 80 countries globally since 1997.
- **Atlantic and Gulf Rapid Reef Assessment (AGRRA)**—Database of 819 survey sites compiled during 39 assessments of the Atlantic region between 1997 and 2004.
- **Global Coral Reef Monitoring Network (GCRMN)**—Regional data on the proportion of coral reef area that experienced bleaching or mortality between 1998 and 2008.
- **Global Environment Monitoring System (GEMS/WATER)**—Data tables of water quality, discharge, and sediment yield statistics for major rivers.
- **MODIS Aqua**—Annual and seasonal composite data of remotely sensed chlorophyll plumes at river mouths.

Expert opinions, combined with the monitoring and assessment data listed above, were used to calibrate the current threat model results. Reef Check and AGRRA data on anthropogenic impacts, coral condition (e.g., live coral cover, algae cover), and species counts were aligned with modeled impacts from overfishing, coastal development, watershed-based pollution, and marine-based pollution. The MODIS Aqua remote sensing data and GEMS/WATER river discharge and sediment yield data were used to calibrate the size of modeled watershed-based pollution plumes. The GCRMN regional bleaching damage and mortality statistics were used to check and calibrate the past thermal stress data layer.

Future thermal stress and ocean acidification data, because of their nature as predictions, could not be calibrated against existing data. As a result, we chose high (conservative) thresholds to establish threat levels.

- **Future thermal stress**—Areas expected to experience a NOAA Bleaching Alert Level 2 at a frequency of 25 percent to 50 percent of years in the decade were classified as medium threat, and areas where the frequency was expected to exceed 50 percent were classified as high threat.
- **Ocean acidification**—Thresholds for aragonite saturation that indicate suitability for coral growth were based on Guinotte, J. M., R. W. Buddemeier, and J. A. Kleypas. 2003. "Future Coral Reef Habitat Marginality: Temporal and Spatial Effects of Climate Change in the Pacific Basin." *Coral Reefs* 22: 551–558. Areas with an aragonite saturation state of 3.25 or greater were classified as under low threat (which is slightly more conservative than a threshold of 3.5 considered "adequate" saturation in Guinotte, et al.); areas between 3.0 and 3.25 were classified as medium threat (considered "low saturation" in Guinotte, et al.), and areas of less than 3.0 were classified as high threat (considered "extremely marginal" in Guinotte, et al.). Furthermore, the CO_2 stabilization levels of 450 ppm and 500 ppm chosen to represent 2030 and 2050, respectively, are slightly more conservative than an IPCC A1B "business-as-usual" emissions scenario in that these CO_2 stabilization levels assume some reduction in global emissions between 2030 and 2050.

INTEGRATED LOCAL THREAT INDEX

In combining the four local threats into the integrated local threat index, we used a method similar to previous *Reefs at Risk* analyses, in that all four threats were weighted equally. The method used, however, is more conservative (i.e., less severe) than previous *Reefs at Risk* analyses, where a "high" in a single threat would put the index to high overall. In the *Reefs at Risk Revisited* modeling, a reef must be threatened by more than one threat to be rated as high threat in the integrated local threat index. The result is a more nuanced distribution of threat among the threat levels. Using this approach, the overall percentage of reefs rated as threatened globally (medium or higher) does not change, and the percentage rated as threatened in any region does not change from the earlier method. The percent rated at high threat is lower than it would be using the traditional method. This new approach prevents the integrated local threat index from being driven too much by a single threat, and provides a more nuanced starting point for the examination of the compound threat associated with future climate change.

INTEGRATED LOCAL THREAT AND FUTURE CLIMATE-RELATED THREAT INDEX

We chose a conservative integration scheme to honor the inherent uncertainties in the projection of future ocean warming and acidification. Using the integrated local threat index as our starting point, the threat level for 2030 or 2050 was only increased if either future thermal stress or acidification threat for the area was rated as high, or both were rated as medium for that time period. If both warming and acidification were rated as high threat, the local threat index is raised two levels. If both acidification and warming were rated as low, or only one was rated as medium, the threat level was not raised. This integration method is intended to be conservative (that is, not alarmist) in looking at the compounded influence of local human pressure and future climate-related threat.

References and Technical Notes

1. Bryant, D., L. Burke, J. McManus, and M. Spalding. 1998. *Reefs at Risk: A Map-Based Indicator of Threats to the World's Coral Reefs*. Washington, DC: World Resources Institute.
2. McAllister, D. 1995. "Status of the World Ocean and Its Biodiversity." *Sea Wind* 9: 1-72.
3. Paulay, G. 1997. "Diversity and Distribution of Reef Organisms." In C. Birkeland, ed. *Life and Death of Coral Reefs*. New York: Chapman & Hall.
4. Calculated at WRI based on data from LandScan High Resolution Global Population Data Set, Oak Ridge National Laboratory, 2007.
5. Jameson, S. C., J. W. McManus, and M. D. Spalding. 1995. *State of the Reefs: Regional and Global Perspectives*. Washington, DC: US Department of State.
6. Jennings, S., and N. V. C. Polunin. 1995. "Comparative Size and Composition of Yield from Six Fijian Reef Fisheries." *Journal of Fish Biology* 46: 28–46.
7. Newton, K., I. M. Côté, G. M. Pilling, S. Jennings, and N. K. Dulvy. 2007. "Current and Future Sustainability of Island Coral Reef Fisheries." *Current Biology* 17: 655–658.
8. The World Bank. 2010. *World Development Indicators*. Accessible at: http://data.worldbank.org/ . Accessed: July 2010.
9. United Nations World Tourism Organization. 2010. *Compendium of Tourism Statistics, Data 2004 - 2008*. Madrid, Spain: World Tourism Organization.
10. U.S. Commission on Ocean Policy. 2004. *An Ocean Blueprint for the 21st Century Final Report*. Washington, DC: U.S. Commission on Ocean Policy.
11. Glaser, K. B., and A. M. S. Mayer. 2009. "A Renaissance in Marine Pharmacology: From Preclinical Curiosity to Clinical Reality." *Biochemical Pharmacology* 78: 440–448.
12. Coastline protected by reefs was calculated at WRI from coastline data from the National Geospatial Intelligence Agency, World Vector Shoreline, 2004; and coral reef data from the Institute for Marine Remote Sensing, University of South Florida (IMaRS/USF), Institut de Recherche pour le Développement (IRD), UNEP-WCMC, The World Fish Center, and WRI, 2011.
13. Fernando, H. J. S., S. P. Samarawickrama, S. Balasubramanian, S. S. L. Hettiarachchi, and S. Voropayev. 2008. "Effects of Porous Barriers Such as Coral Reefs on Coastal Wave Propagation." *Journal of Hydro-environment Research* 1: 187–194; Sheppard, C., D. J. Dixon, M. Gourlay, A. Sheppard, and R. Payet.2005. "Coral Mortality Increases Wave Energy Reaching Shores Protected by Reef Flats: Examples from the Seychelles." *Estuarine, Coastal and Shelf Science* 64: 223–234.
14. Spalding, M., C. Ravilious, and E. P. Green. 2001. *World Atlas of Coral Reefs*. Berkeley, CA: University of California Press.
15. The coral reef data used in the Reefs at Risk Revisited analysis were compiled specifically for this project from multiple sources by UNEP-WCMC, the World Fish Center, and WRI, incorporating products from the Millennium Coral Reef Mapping Project prepared by the Institute for Marine Remote Sensing, University of South Florida (IMaRS/USF), Institut de Recherche pour le Développement (IRD), 2011. To standardize these data for the purposes of the Reefs at Risk Revisited project, data were converted to raster format (ESRI grid) at 500-m resolution.
16. Sadovy, Y. 2005. "Trouble on the Reef: The Imperative for Managing Valuable and Vulnerable Fisheries." *Fish and Fisheries* 6: 167–185.
17. Jackson, J. B. C. 2008. "Ecological Extinction and Evolution in the Brave New Ocean." *Proceedings of the National Academy of Sciences* 105: 11458–11465.
18. Mumby, P. J. et al. 2006. "Fishing, Trophic Cascades, and the Process of Grazing on Coral Reefs." *Science* 311: 98–101.
19. Roberts, C. M. 1995. "Effects of Fishing on the Ecosystem Structure of Coral Reefs." *Conservation Biology* 9: 988–995.
20. Silverman, J., B. Lazar, L. Cao, K. Caldeira, and J. Erez. 2009. "Coral Reefs May Start Dissolving When Atmospheric CO_2 Doubles." *Geophysical Research Letters* 36: L05606.
21. Hughes, T. P., M. J. Rodrigues, D. R. Bellwood, D. Ceccarelli, O. Hoegh-Guldberg, L. McCook, N. Moltschaniwskyj, M. S. Pratchett, R. S. Steneck, and B. Willis.2007. "Phase Shifts, Herbivory, and the Resilience of Coral Reefs to Climate Change." *Current Biology* 17: 360–365.
22. Riegl, B., A. Bruckner, S. L. Coles, P. Renaud, and R. E. Dodge. 2009. "Coral Reefs: Threats and Conservation in an Era of Global Change." *Annals of the New York Academy of Sciences* 1162: 136–186.
23. Threat, in this analysis, is defined as a level of human use or influence that has the potential to drive major declines in natural ecosystem function or in the provision of ecosystem services to local populations within 10 years. We consider major declines to be large and long-term changes to ecological function, biodiversity, biomass, or productivity, and to be distinct from the minor detectable changes that are already widespread on almost every reef. The level of threat gives an indication of likelihood of occurrence or of the severity of potential impact, or both.
24. Although this combined map gives a better assessment of the present day threats to reefs at the global scale, the low spatial accuracy of the past thermal stress model means that it is less valuable for detailed spatial analysis and investigation.
25. Calculated at WRI based on data from LandScan High Resolution Global Population Data Set, Oak Ridge National Laboratory, 2007.
26. Reopanichkul, P., T. A. Schlacher, R. W. Carter, and S. Worachananant. 2009. "Sewage Impacts Coral Reefs at Multiple Levels of Ecological Organization." *Marine Pollution Bulletin* 58: 1356–1362; Pastorok, R. A., and G. R. Bilyard. 1985. "Effects of Sewage Pollution on Coral Reef Communities." *Marine Ecology Progress Series* 21: 175–189.

27. Jeftic, L. and United Nations Environment Programme. 2006. *The State of the Marine Environment : Trends and Processes*. The Hague: UNEP/GPA Coordination Office.

28. Hawkins, J. P., and C. M. Roberts.1994. "The Growth of Coastal Tourism in the Red Sea: Present and Future Effects on Coral Reefs." *Ambio* 23: 503–508.

29. Calculated at WRI from 2000 and 2005 population density data from the Center for International Earth Science Information Network (CIESIN), Columbia University; and the Global Rural Urban Mapping Project (GRUMP) Alpha and Beta Versions, Columbia University.

30. Clark, J. 1997. "Coastal Zone Management for the New Century." *Ocean and Coastal Management* 37: 191–216.

31. Coastal Zone Management Unit. 2006. *Integrated Coastal Management Plan*. St. Michael: Barbados.

32. Rogers, C. S. 1990. "Responses of Coral Reefs and Reef Organisms to Sedimentation." *Marine ecology progress series. Oldendorf* 62: 185–202.

33. Alongi, D. M., and A. D. McKinnon. 2005. "The Cycling and Fate of Terrestrially-Derived Sediments and Nutrients in the Coastal Zone of the Great Barrier Reef Shelf." *Marine Pollution Bulletin* 51: 239–252; Nedwell, D. B. 1975. "Inorganic Nitrogen Metabolism in a Eutrophicated Tropical Mangrove Estuary." *Water Research* 9: 221–231.

34. Spalding, M. D., M. Kainuma, and L. Collins. 2010. *World Atlas of Mangroves*. London: Earthscan, with International Society for Mangrove Ecosystems, Food and Agriculture Organization of the United Nations, UNEP World Conservation Monitoring Centre, United Nations Scientific and Cultural Organisation, and United Nations University.

35. Mumby, P. J. 2006. "Connectivity of Reef Fish between Mangroves and Coral Reefs: Algorithms for the Design of Marine Reserves at Seascape Scales." *Biological Conservation* 128: 215–222; Nagelkerken, I., S. J. M. Blaber, S. Bouillon, P. Green, M. Haywood, L. G. Kirton, J. O. Meynecke, J. Pawlik, H. M. Penrose, A. Sasekumar, and P. J. Somerfield. 2008. "The Habitat Function of Mangroves for Terrestrial and Marine Fauna: A Review." *Aquatic Botany* 89: 155–185.

36. FAO.2000. *Fertilizer Requirements in 2015 and 2030*. Rome: Food and Agriculture Organization of the United Nations.

37. Crossland, C. J., H. H. Kremer, H. J. Lindeboom, J. I. M. Crossland, and M. D. A. Le Tissier. 2005. *Coastal Fluxes in the Anthropocene: The Land-Ocean Interactions in the Coastal Zone Project of the International Geosphere-Biosphere Programme*. Heidelberg: Springer Verlag.

38. Wright, L. D., and C. A. Nittrouer. 1995. "Dispersal of River Sediments in Coastal Seas: Six Contrasting Cases." *Estuaries and Coasts* 18: 494–508.

39. Waycott, M. et al. 2009. "Accelerating Loss of Seagrasses across the Globe Threatens Coastal Ecosystems." *Proceedings of the National Academy of Sciences* 106: 12377–12381.

40. Buddemeier, R., J. Kleypas, and R. Aronson. 2004. *Coral Reefs and Global Climate Change: Potential Contributions of Climate Change to Stresses on Coral Reef Ecosystems*. Arlington, VA: Pew Center on Global Climate Change.

41. Selman, M., S. Greenhalgh, R. Diaz, and Z. Sugg.2008. Eutrophication and Hypoxia in Coastal Areas: A Global Assessment of the State of Knowledge. Washington, DC: World Resources Institute.

42. Hansen, M. C. et al. 2008. "Humid Tropical Forest Clearing from 2000 to 2005 Quantified by Using Multitemporal and Multiresolution Remotely Sensed Data." *Proceedings of the National Academy of Sciences* 105: 9439–9444; Forestry Department, FAO. 2006.*Global Forest Resources Assessment 2005: Progress Towards Sustainable Forest Management*. Rome: FAO.

43. Trenberth, K. E. et al. 2007. "Observations: Surface and Atmospheric Climate Change." In S. Solomon et al., eds. *Climate Change 2007: The Physical Science Basis. Contribution of Working Group I to the Fourth Assessment Report of the Intergovernmental Panel on Climate Change*. Cambridge, UK: Cambridge University Press.

44. Hoff, R. Z., and G. Shigenaka. 2001. *Oil Spills in Coral Reefs: Planning and Response Considerations*. Silver Spring, MD: NOAA, National Ocean Service, Office of Response and Restoration.

45. Cruise Lines International Association Inc. 2010. *The State of the Cruise Industry in 2010: Confident and Offering New Ships, Innovation, and Exception Value*. Accessible at:www.cruising.org/vacation/news/press_releases/2010/01/state-cruise-industry-2010-confident-and-offering-new-ships-innovation. Accessed: July 2010.

46. Ruiz, G. M., J. T. Carlton, E. D. Grosholz, and A. H. Hines. 1997. "Global Invasions of Marine and Estuarine Habitats by Non-Indigenous Species: Mechanisms, Extent, and Consequences." *Integrative and Comparative Biology* 37: 621–632; Molnar, J. L., R. L. Gamboa, C. Revenga, and M. D. Spalding. 2008. "Assessing the Global Threat of Invasive Species to Marine Biodiversity." *Frontiers in Ecology and the Environment* 6: 485–492.

47. Wilkinson, C. 2008. *Status of Coral Reefs of the World: 2008*. Townsville, Australia: Global Coral Reef Monitoring Network and Reef and Rainforest Research Centre.

48. Carton, J. T. 1999. "The Scale and Ecological Consequences of Biological Invasions in the World's Oceans." In O.T. Sandlund, P.J. Schei, and A. Viken, eds. *Invasive Species and Biodiversity Management*. Dordrecht, The Netherlands: Kluwer Academic Publishers.

49. Andrews, K., L. Nall, C. Jeffrey, S. Pittman, K. Banks, C. Beaver, J. Bohnsack, R. E. Dodge, D. Gilliam, W. Jaap, and others. 2005. "The State of Coral Reef Ecosystems of Florida." In J. E. Waddell and A. M. Clarke, eds. *The State of Coral Reef Ecosystems of the United States and Pacific Freely Associated States*. Silver Spring, MD: NOAA, National Centers for Coastal Ocean Science.

50. Friedlander, A., G. Aeby, E. Brown, A. Clark, S. Coles, S. Dollar, C. Hunter, P. Jokiel, J. Smith, B. Walsh, and others. 2008. "The State of Coral Reef Ecosystems of the Main Hawaiian Islands." In J. E. Waddell and A. M. Clarke, eds. *The State of Coral Reef Ecosystems of the United States and Pacific Freely Associated States*. Silver Spring, MD: NOAA, National Centers for Coastal Ocean Science.

51. Jaap, W. C. 2000. "Coral Reef Restoration." *Ecological Engineering* 15: 345–364.

52. U.S. Energy Information Administration. 2007. *International Energy Outlook 2007.* Washington, DC: US Department of Energy.

53. Crone, T. J., and M. Tolstoy.2010. "Magnitude of the 2010 Gulf of Mexico Oil Leak." *Science* 330: 634–634; National Academy of Engineering and National Research Council. 2010. *Interim Report on Causes of the Deepwater Horizon Oil Rig Blowout and Ways to Prevent Such Events.* Washington, DC: National Academy of Engineering and National Research Council.

54. Christiansen, M., K. Fagerholt, B. R. Nygreen, D. Ronen, B. Cynthia, and L. Gilbert. 2007. "Maritime Transportation." In C. Barnhart and G. Laporte, eds. *Transportation.* Volume 14. Amsterdam: Elsevier.

55. Sweeting Sweeting, J., and S. Wayne. 2003. *A Shifting Tide: Environmental Challenges and Cruise Industry Responses.* Washington, DC: Conservation International, The Center for Environmental Leadership in Business.

56. Calculated at WRI based on data from LandScan High Resolution Global Population Data Set, Oak Ridge National Laboratory, 2007.

57. Graham, N. A. J., M. Spalding, and C. Sheppard. 2010. "Reef Shark Declines in Remote Atolls Highlight the Need for Multi-Faceted Conservation Action." *Aquatic Conservation: Marine and Freshwater Ecosystems* 20: 543–548.

58. Berkes, F. et al.2006. "Globalization, Roving Bandits, and Marine Resources." *Science* 311: 1557–1558; Sadovy, Y. J., T. J. Donaldson, T. R. Graham, F. McGilvray, G. J. Muldoon, M. J. Phillips, M. A. Rimmer, A. Smith, and B. Yeeting.2003. *While Stocks Last: The Live Reef Food Fish Trade.* Manila: Asian Development Bank.

59. Burkepile, D. E., and M. E. Hay. 2008. "Herbivore Species Richness and Feeding Complementarity Affect Community Structure and Function on a Coral Reef." *Proceedings of the National Academy of Sciences* 105: 16201–16206; Friedlander, A. M., and E. E. DeMartini. 2002. "Contrasts in Density, Size, and Biomass of Reef Fishes between the Northwestern and the Main Hawaiian Islands: The Effects of Fishing Down Apex Predators." *Marine Ecology Progress Series* 230: 253–264; Pauly, D., V. Christensen, J. Dalsgaard, R. Froese, and F. Torres. 1998. "Fishing Down Marine Food Webs." *Science* 279: 860–863.

60. Hughes, T. P., D. R. Bellwood, C. S. Folke, L. J. McCook, and J. M. Pandolfi. 2007. "No-Take Areas, Herbivory and Coral Reef Resilience." *Trends in Ecology and Evolution* 22: 1–3.

61. Mumby, P. J., and A. R. Harborne. 2010. "Marine Reserves Enhance the Recovery of Corals on Caribbean Reefs." *PLoS ONE* 5: e8657.

62. Raymundo, L. J., A. R. Halford, A. P. Maypa, and A. M. Kerr.2009. "Functionally Diverse Reef-Fish Communities Ameliorate Coral Disease." *Proceedings of the National Academy of Sciences* 106: 17067–17070.

63. Fox, H. E., and R. L. Caldwell. 2006. "Recovery from Blast Fishing on Coral Reefs: A Tale of Two Scales." *Ecological Applications* 16: 1631–1635.

64. Fox, H. E., J. S. Pet, R. Dahuri, R. L. Caldwell, M. K. Moosa, S. Soemodihardjo, A. Soegiarto, K. Romimohtarto, A. Nontji, and S. Suharsono. 2002. *Coral Reef Restoration after Blast Fishing in Indonesia.* Bali, Indonesia: Proceedings of the 9th International Coral Reef Symposium; Wells, S. 2009. "Dynamite Fishing in Northern Tanzania–Pervasive, Problematic and yet Preventable." *Marine Pollution Bulletin* 58: 20–23.

65. Mous, P. J., L. Pet-Soede, M. Erdmann, H. S. J. Cesar, Y. Sadovy, and J. Pet. 2000. "Cyanide Fishing on Indonesian Coral Reefs for the Live Food Fish Market—What Is the Problem?" *SPC Live Reef Fish Information Bulletin* 7: 20–27; Barber, C. V., and V. R. Pratt. 1997. *Sullied Seas: Strategies for Combating Cyanide Fishing in Southeast Asia.* Washington, DC: World Resources Institute.

66. FAO Fisheries Aquaculture Dept. 2009. *The State of World Fisheries and Aquaculture: 2008.* Rome: FAO.

67. Agnew, D. J., J. Pearce, G. Pramod, T. Peatman, R. Watson, J. R. Beddington, and T. J. Pitcher. 2009."Estimating the Worldwide Extent of Illegal Fishing." *PLoS ONE* 4: e4570.

68. FAO and WorldFish Center. 2008. *Small-Scale Capture Fisheries: A Global Overview with Emphasis on Developing Countries.* A Preliminary Report of the Big Numbers Project. Penang, Malaysia: World Fish Center.

69. Zeller, D., S. Booth, P. Craig, and D. Pauly. 2005. "Reconstruction of Coral Reef Fisheries Catches in American Samoa, 1950–2002." *Coral Reefs* 25: 144–152.

70. Friedlander, A., G. Aeby, S. Balwani, B. Bowen, R. Brainard, A. Clark, J. Kenyon, J. Maragos, C. Meyer, P. Vroom, and J. Zamzow. 2008. "The State of Coral Reef Ecosystems of the Northwestern Hawaiian Islands." In J. E. Waddell and A. M. Clarke, eds. *The State of Coral Reef Ecosystems of the United States and Pacific Freely Associated States.* Silver Spring, MD: NOAA, National Centers for Coastal Ocean Science.

71. The government of the United Kingdom declared the Chagos Archipelago Marine Protected Area in April 2010. The exact boundaries and regulations were not in place at the time of publication, though all commercial fishing was formally ended on November 1, 2010.

72. Bellwood, D. R., T. P. Hughes, C. Folke, and M. Nyström. 2004. "Confronting the Coral Reef Crisis." *Nature* 429: 827–833.

73. Great Barrier Reef Marine Park Authority. 2009. *Great Barrier Reef Outlook Report 2009.* Townsville, Australia: Great Barrier Reef Marine Park Authority.

74. Russ, G. R., A. J. Cheal, A. M. Dolman, M. J. Emslie, R. D. Evans, I. Miller, H. Sweatman, and D. H. Williamson. 2008. "Rapid Increase in Fish Numbers Follows Creation of World's Largest Marine Reserve Network." *Current Biology* 18: 514–515.

75. Sweatman, H. 2008. "No-Take Reserves Protect Coral Reefs from Predatory Starfish." *Current Biology* 18: R598–599.

76. Wabnitz, C., M. Taylor, E. Green, and T. Razak.2003. *From Ocean to Aquarium : The Global Trade in Marine Ornamental Species.* Cambridge, UK: UNEP World Conservation Monitoring Centre.

77. Erdmann, M. V., and L. Pet-Soede. 1996. "How Fresh Is Too Fresh? The Live Reef Food Fish Trade in Eastern Indonesia." *NAGA, the ICLARM quarterly* 19: 4–8.

78. Sadovy, Y. J., T. J. Donaldson, T. R. Graham, F. McGilvray, G. J. Muldoon, M. J. Phillips, M. A. Rimmer, A. Smith and B. Yeeting. *While Stocks Last: The Live Reef Food Fish Trade.* (Asian Development Bank, 2003).

79. McClanahan, T. R., C. C. Hicks, and E. S. Darling. 2008. "Malthusian Overfishing and Efforts to Overcome It on Kenyan Coral Reefs." *Ecological Applications* 18: 1516–1529; Hilborn, R. 2007. "Moving to Sustainability by Learning from Successful Fisheries." *AMBIO: A Journal of the Human Environment* 36: 296–303.

80. Cinner, J. E., T. R. McClanahan, T. M. Daw, N. A. J. Graham, J. Maina, S. K. Wilson, and T. P. Hughes. 2009. "Linking Social and Ecological Systems to Sustain Coral Reef Fisheries." *Current Biology* 19: 206–212; Alcala, A. C., G. R. Russ, A. P. Maypa, and H. P. Calumpong. 2005." A Long-Term, Spatially Replicated Experimental Test of the Effect of Marine Reserves on Local Fish Yields." *Canadian Journal of Fisheries and Aquatic Sciences* 62: 98–108.

81. Lester, S., and B. Halpern. 2008. "Biological Responses in Marine No-Take Reserves Versus Partially Protected Areas." *Marine Ecology Progress Series* 367: 49–56.

82. McClellan, K., and J. Bruno.2008. *Coral Degradation through Destructive Fishing Practices - Encyclopedia of Earth.* Washington, DC: Environmental Information Coalition, National Council for Science and the Environment.

83. Gulbrandsen, L. H. 2009. "The Emergence and Effectiveness of the Marine Stewardship Council." *Marine Policy* 33: 654–660; Jacquet, J., J. Hocevar, S. Lai, P. Majluf, N. Pelletier, T. Pitcher, E. Sala, R. Sumaila, and D. Pauly. 2009. "Conserving Wild Fish in a Sea of Market-Based Efforts." *Oryx* 44: 45–56; Leadbitter, D., G. Gomez, and F. McGilvray. 2006. "Sustainable Fisheries and the East Asian Seas: Can the Private Sector Play a Role?" *Ocean and Coastal Management* 49: 662–675; Shelton, P. A. 2009. "Eco-Certification of Sustainably Managed Fisheries--Redundancy or Synergy?" *Fisheries Research* 100: 185–190; Ward, T. J. 2008. "Barriers to Biodiversity Conservation in Marine Fishery Certification." *Fish and Fisheries* 9: 169–177.

84. Eakin, C. M., J. M. Lough, and S. F. Heron. 2009. "Climate Variability and Change: Monitoring Data and Evidence for Increased Coral Bleaching Stress." In M. J. H. Oppen and J. M. Lough, eds. *Coral Bleaching.* Vol. 205, *Ecological Studies.* Heidelberg, Germany: Springer.

85. Glynn, P. W. 1993. "Coral Reef Bleaching: Ecological Perspectives." *Coral Reefs* 12: 1–17.

86. Hoegh-Guldberg, O. 1999. "Climate Change, Coral Bleaching and the Future of the World's Coral Reefs." *Marine and Freshwater Research* 50: 839–866.

87. Reaser, J. K., R. Pomerance, and P. O. Thomas. 2000. "Coral Bleaching and Global Climate Change: Scientific Findings and Policy Recommendations." *Conservation Biology* 14: 1500–1511.

88. Oliver, J. K., R. Berkelmans, and C. M. Eakin. 2009. "Coral Bleaching in Space and Time." In M. J. H. Oppen and J. M. Lough, eds. *Coral Bleaching.* Vol. 205, *Ecological Studies.* Heidelberg, Germany: Springer.

89. Goreau, T., T. McClanahan, R. Hayes, and A. Strong. 2000. "Conservation of Coral Reefs after the 1998 Global Bleaching Event." *Conservation Biology* 14: 5–15; Lindén, O., and N. Sporrong. 1999. *Coral Reef Degradation in the Indian Ocean: Status Reports and Project Presentations 1999.* Stockholm: Stockholm University, CORDIO; Souter, D., D. Obura and O. Lindén. 2000. *Coral Reef Degradation in the Indian Ocean. Status Report 2000.* Stockholmn, Sweden: CORDIO, SAREC Marine Science Program.

90. Sheppard, C. R. C., M. Spalding, C. Bradshaw, and S. Wilson. 2002. "Erosion vs. Recovery of Coral Reefs after 1998 El Niño: Chagos Reefs, Indian Ocean." *Ambio* 31: 40–48.

91. Berkelmans, R., G. De'ath, S. Kininmonth, and W. J. Skirving. 2004. "A Comparison of the 1998 and 2002 Coral Bleaching Events on the Great Barrier Reef: Spatial Correlation, Patterns, and Predictions." *Coral Reefs* 23: 74–83.

92. Wilkinson, C., and D. Souter.2008. *Status of Caribbean Coral Reefs after Bleaching and Hurricanes in 2005.* Townsville, Australia: Global Coral Reef Monitoring Network, and Reef and Rainforest Research Centre.

93. Eakin, C. M. et al. 2010. "Caribbean Corals in Crisis: Record Thermal Stress, Bleaching, and Mortality in 2005." *PLoS ONE* 5: e13969.

94. Veron, J. E. N., O. Hoegh-Guldberg, T. M. Lenton, J. M. Lough, D. O. Obura, P. Pearce-Kelly, C. R. C. Sheppard, M. Spalding, M. G. Stafford-Smith, and A. D. Rogers. 2009. "The Coral Reef Crisis: The Critical Importance of< 350ppm CO_2." *Marine Pollution Bulletin* 58: 1428–1436.

95. Baird, A., and J. Maynard.2008. "Coral Adaptation in the Face of Climate Change." *Science* 320: 315; Maynard, J. A., K. R. N. Anthony, P. A. Marshall, and I. Masiri. 2008. "Major Bleaching Events Can Lead to Increased Thermal Tolerance in Corals." *Marine Biology* 155: 173–182.

96. Obura, D., N. 2009. *Resilience Assessment of Coral Reefs Assessment Protocol for Coral Reefs, Focusing on Coral Bleaching and Thermal Stress.* Gland, Switzerland: IUCN.

97. Grimsditch, G. D., and R. V. Salm. 2006. *Coral Reef Resilience and Resistance to Bleaching.* Gland, Switzerland: IUCN.

98. Lasagna, R., G. Albertelli, P. Colantoni, C. Morri, and C. Bianchi. 2009. "Ecological Stages of Maldivian Reefs after the Coral Mass Mortality of 1998." *Facies* 56: 1–11.

99. Sheppard, C. R. C., A. Harris, and A. L. S. Sheppard. 2008. "Archipelago-Wide Coral Recovery Patterns since 1998 in the Chagos Archipelago, Central Indian Ocean." *Marine Ecology Progress Series* 362: 109–117.

100. Graham, N. A. J., S. K. Wilson, S. Jennings, N. V. C. Polunin, J. P. Bijoux, and J. Robinson. 2006. "Dynamic Fragility of Oceanic Coral Reef Ecosystems." *Proceedings of the National Academy of Sciences* 103: 8425–8429.

101. Obura, D. 2005. "Resilience and Climate Change: Lessons from Coral Reefs and Bleaching in the Western Indian Ocean." *Estuarine, Coastal and Shelf Science* 63: 353–372.

102. Baker, A. C., P. W. Glynn, and B. Riegl.2008. "Climate Change and Coral Reef Bleaching: An Ecological Assessment of Long-Term Impacts, Recovery Trends and Future Outlook." *Estuarine, Coastal and Shelf Science* 80: 435–471.

103. Coelho, V. R., and C. Manfrino. 2007. "Coral Community Decline at a Remote Caribbean Island: Marine No-Take Reserves Are Not Enough." *Aquatic Conservation: Marine and Freshwater Ecosystems 17*: 666–685.

104. Somerfield, P., W. Jaap, K. Clarke, M. Callahan, K. Hackett, J. Porter, M. Lybolt, C. Tsokos, and G. Yanev. 2008. "Changes in Coral Reef Communities among the Florida Keys, 1996–2003." *Coral Reefs* 27: 951–965.

105. Carilli, J., R. Norris, B. Black, S. Walsh, and M. McField. 2009. "Local Stressors Reduce Coral Resilience to Bleaching." *PLoS One* 4: e6324.

106. IUCN. 2010. *IUCN Red List of Threatened Species. Version 2010.4*. Accessible at: www.iucnredlist.org . Accessed: November 22, 2010.

107. Carpenter, K. E. et al. 2008. "One-Third of Reef-Building Corals Face Elevated Extinction Risk from Climate Change and Local Impacts." *Science* 321: 560–563.

108. Hawkins, J. P., C. M. Roberts, and V. Clark. 2000. "The Threatened Status of Restricted-Range Coral Reef Fish Species." *Animal Conservation* 3: 81–88.

109. Strong, A. E., F. Arzayus, W. Skirving, and S. F. Heron.2006."Identifying Coral Bleaching Remotely Via Coral Reef Watch - Improved Integration and Implications for Changing Climate." In J. T. Phinney, et al., eds. *Coral Reefs and Climate Change: Science and Management.* Vol. 61, *Coastal and Estuarine Studies*. Washington, DC: American Geophysical Union.

110. Casey, K. S., T. B. Brandon, P. Cornillon, and R. Evans. 2010. "The Past, Present and Future of the AVHRR Pathfinder SST Program." In V. Barale, J. F. R. Gower, and L. Alberotanza, eds. *Oceanography from Space: Revisited.* New York: Springer.

111. A higher DHW threshold was used for the Middle East region (Red Sea and Persian Gulf) to compensate for exaggerated temperature readings driven by land around these enclosed seas. See the full technical notes at www.wri.org/reefs for detailed information on the modification and justification.

112. The bleaching observations from the ReefBase database (or any other compilation) are spatially and temporally limited due to a lack of oberservers in remote locations and limited reporting effort (that is, observations often go unreported). Therefore, we used the satellite-detected thermal stress as a means of filling in the gaps in observational data.

113. Lough, J. M. 2000. "1997-98: Unprecedented Thermal Stress to Coral Reefs?" *Geophysical Research Letters* 27: 3901–3904; McWilliams, J. P., I. M. Côté, J. A. Gill, W. J. Sutherland, and A. R. Watkinson. 2005. "Accelerating Impacts of Temperature-Induced Coral Bleaching in the Caribbean." *Ecology* 86: 2055–2060.

114. Sheppard, C. R. C. 2003. "Predicted Recurrences of Mass Coral Mortality in the Indian Ocean." *Nature* 425: 294–297.

115. Donner, S. D., W. J. Skirving, C. M. Little, M. Oppenheimer, and O. Hoegh-Guldberg.2005. "Global Assessment of Coral Bleaching and Required Rates of Adaptation under Climate Change." *Global Change Biology* 11: 2251–2265.

116. Canadell, J., P. Ciais, D. S, C. Le Quéré, A. Patwardhan, and M. Raupach. 2009. *The Human Perturbation of the Carbon Cycle.* Paris: UNESCO-SCOPE-UNEP.

117. Donner, S. 2009. "Coping with Commitment: Projected Thermal Stress on Coral Reefs under Different Future Scenarios." *PLoS ONE* 4: e5712.

118. Marshall, P., and H. Schuttenberg. 2006. *A Reef Manager's Guide to Coral Bleaching*. Townsville, Australia: Great Barrier Reef Marine Park Authority.

119. West, J. M., and R. V. Salm. 2003. "Resistance and Resilience to Coral Bleaching: Implications for Coral Reef Conservation and Management." *Conservation Biology* 17: 956–967.

120. Anthony, K. R. N., D. I. Kline, G. Diaz-Pulido, S. Dove, and O. Hoegh-Guldberg. 2008. "Ocean Acidification Causes Bleaching and Productivity Loss in Coral Reef Builders." *Proceedings of the National Academy of Sciences* 105: 17442.

121. Sabine, C. L. 2004. "The Oceanic Sink for Anthropogenic CO_2." *Science* 305: 367–371.

122. Cao, L., K. Caldeira, and A. K. Jain. 2007. "Effects of Carbon Dioxide and Climate Change on Ocean Acidification and Carbonate Mineral Saturation." *Geophysical Research Letters* 34: 5607; Guinotte, J. M., and V. J. Fabry. 2008. "Ocean Acidification and Its Potential Effects on Marine Ecosystems." *Annals of the New York Academy of Sciences* 1134: 320–342; Kuffner, I. B., A. J. Andersson, P. L. Jokiel, K. u. S. Rodgers, and F. T. Mackenzie. 2008. "Decreased Abundance of Crustose Coralline Algae Due to Ocean Acidification." *Nature Geoscience* 1: 114–117.

123. Guinotte, J. M., R. W. Buddemeier, and J. A. Kleypas. 2003. "Future Coral Reef Habitat Marginality: Temporal and Spatial Effects of Climate Change in the Pacific Basin." *Coral Reefs* 22: 551–558.

124. Bak, R. P. M., G. Nieuwland, and E. H. Meesters. 2009. "Coral Growth Rates Revisited after 31 Years: What Is Causing Lower Extension Rates in Acropora Palmata?" *Bulletin of Marine Science* 84: 287–294; Cooper, T. F., G. De'Ath, K. E. Fabricius, and J. M. Lough. 2008. "Declining Coral Calcification in Massive Porites in Two Nearshore Regions of the Northern Great Barrier Reef." *Global Change Biology* 14: 529–538; De'ath, G., J. M. Lough, and K. E. Fabricius. 2009."Declining Coral Calcification on the Great Barrier Reef." *Science* 323: 116–119.

125. Kleypas, J. A., R. W. Buddemeier, D. Archer, J. P. Gattuso, C. Langdon, and B. N. Opdyke. 1999. "Geochemical Consequences of Increased Atmospheric Carbon Dioxide on Coral Reefs." *Science* 284: 118.

126. Hoegh-Guldberg, O., et al. 2007. "Coral Reefs under Rapid Climate Change and Ocean Acidification." *Science* 318: 1737–1742.

127. Veron, J. 2008. "Mass Extinctions and Ocean Acidification: Biological Constraints on Geological Dilemmas." *Coral Reefs* 27: 459–472.

128. Hansen, J., M. Sato, P. Kharecha, D. Beerling, R. Berner, V. Masson-Delmotte, M. Pagani, M. Raymo, D. L. Royer, and J. C. Zachos. "Target Atmospheric CO_2: Where Should Humanity Aim?" *The Open Atmospheric Science Journal* 2: 217–231.

129. Allison, I., et al. 2009. *The Copenhagen Diagnosis, 2009: Updating the World on the Latest Climate Science*. Sydney, Australia: University of New South Wales, Climate Change Research Centre.

130. IPCC. 2007. *Synthesis Report.* Contribution of Working Groups I, II and III to the Fourth Assessment Report of the Intergovernmental Panel on Climate Change. Geneva: Intergovernmental Panel on Climate Change.

131. Grinsted, A., J. Moore, and S. Jevrejeva. 2009."Reconstructing Sea Level from Paleo and Projected Temperatures 200 to 2100 AD." *Climate Dynamics* 10: 461-472.

132. Webb, A., and P. Kench. 2010. "The Dynamic Response of Reef Islands to Sea-Level Rise: Evidence from Multi-Decadal Analysis of Island Change in the Central Pacific." *Global and Planetary Change* 72: 234–246.

133. Pilkey, O. H., and J. A. G. Cooper. 2004. "Society and Sea Level Rise." *Science* 303: 1781–1782.

134. Woodroffe, C. D. 2008. "Reef-Island Topography and the Vulnerability of Atolls to Sea-Level Rise." *Global and Planetary Change* 62: 77–96.

135. Ulbrich, U., G. Leckebusch, and J. Pinto.2009. "Extra-Tropical Cyclones in the Present and Future Climate: A Review." *Theoretical and Applied Climatology* 96: 117–131.

136. Emanuel, K., R. Sundararajan, and J. Williams. 2008. "Hurricanes and Global Warming: Results from Downscaling IPCC Ar4 Simulations."*American Meteorological Society* (March 2008): 347–367.

137. Sutherland, K. P., J. W. Porter, and C. Torres. 2004. "Disease and Immunity in Caribbean and Indo-Pacific Zooxanthellate Corals." *Marine Ecology Progress Series* 266: 273–302.

138. Harvell, C. D., and E. Jordán-Dahlgren. 2007. "Coral Disease, Environmental Drivers, and the Balance between Coral and Microbial Associates." *Oceanography and Marine Biology: an Annual Review* 20: 58–81.

139. Bruno, J. F., E. R. Selig, K. S. Casey, C. A. Page, B. L. Willis, C. D. Harvell, H. Sweatman, and A. M. Melendy. 2007. "Thermal Stress and Coral Cover as Drivers of Coral Disease Outbreaks." *PLoS Biol* 5: 1220–1227.

140. Gardner, T. A., I. M. Cote, J. A. Gill, A. Grant, and A. R. Watkinson. 2003. "Long-Term Region-Wide Declines in Caribbean Corals." *Science* 301: 958–960.

141. Aronson, R. B., and W. F. Precht. 2002. "White-Band Disease and the Changing Face of Caribbean Coral Reefs." *Hydrobiologia* 460: 25–38.

142. Lessios, H. 1988. "Mass Mortality of Diadema Antillarum in the Caribbean: What Have We Learned?" *Annual Review of Ecology, Evolution, and Systematics* 19: 371–393.

143. Edmunds, P., and R. Carpenter. 2001. "Recovery of Diadema Antillarum Reduces Macroalgal Cover and Increases Abundance of Juvenile Corals on a Caribbean Reef." *Proc Natl Acad Sci USA* 98: 5067–5071; Idjadi, J., R. Haring, and W. Precht. 2010. "Recovery of the Sea Urchin Diadema Antillarum Promotes Scleractinian Coral Growth and Survivorship on Shallow Jamaican Reefs." *Marine Ecology Progress Series* 403: 91–100.

144. Raymundo, L. J., C. S. Couch, and C. D. Harvell. 2008. *Coral Disease Handbook: Guidelines for Assessment, Monitoring and Management.* Melbourne, Australia: Coral Reef Targeted Research and Capacity Building for Management Program.

145. Lesser, M. P., J. C. Bythell, R. D. Gates, R. W. Johnstone, and O. Hoegh-Guldberg. 2007. "Are Infectious Diseases Really Killing Corals? Alternative Interpretations of the Experimental and Ecological Data." *Journal of Experimental Marine Biology and Ecology* 346: 36–44.

146. Sapp, J. 1999. *What Is Natural? Coral Reef Crisis.* Oxford: Oxford University Press.

147. Dulvy, N. K., R. P. Freckleton, and N. V. C. Polunin. 2004. "Coral Reef Cascades and the Indirect Effects of Predator Removal by Exploitation." *Ecology Letters* 7: 410–416.

148. Brodie, J., K. Fabricius, G. De'ath, and K. Okaji. 2005. "Are Increased Nutrient Inputs Responsible for More Outbreaks of Crown-of-Thorns Starfish? An Appraisal of the Evidence." *Marine Pollution Bulletin* 51: 266.

149. McClanahan, T. R., M. Ateweberhan, C. R. Sebastián, N. A. J. Graham, S. K. Wilson, J. H. Bruggemann, and M. M. M. Guillaume. 2007. "Predictability of Coral Bleaching from Synoptic Satellite and in Situ Temperature Observations." *Coral Reefs* 26: 695–701.

150. Maina, J., V. Venus, T. R. McClanahan, and M. Ateweberhan. 2008. "Modelling Susceptibility of Coral Reefs to Environmental Stress Using Remote Sensing Data and Gis Models." *Ecological Modelling* 212: 180–199.

151. Kleypas, J. A., G. Danabasoglu, and J. M. Lough. 2008. "Potential Role of the Ocean Thermostat in Determining Regional Differences in Coral Reef Bleaching Events." *Geophysical Research Letters* 35: L03613.

152. Grimsditch, G. 2006. *Coral Reef Resilience and Resistance to Bleaching.* Gland, Switzerland: IUCN, The World Conservation Union.

153. United Nations. 2004. *World Population to 2300.* New York: United Nations.

154. Manzello, D. 2010. "Coral Growth with Thermal Stress and Ocean Acidification: Lessons from the Eastern Tropical Pacific." *Coral Reefs* 29: 749–758.

155. Randall, J. E. 1998. "Zoogeography of Shore Fishes of the Indo-Pacific Region." *Zoological Studies* 37: 227–268.

156. U.S. Energy Information Administration. 2008. *World Oil Transit Chokepoints: Strait of Hormuz.* Accessible at: www.eia.doe.gov/cabs/World_Oil_Transit_Chokepoints/Hormuz.html. Accessed: September 6, 2010.

157. Calculated at WRI based on data from LandScan High Resolution Global Population Data Set, Oak Ridge National Laboratory, 2007 and coral reef data from the Institute for Marine Remote Sensing, University of South Florida (IMaRS/USF), Institut de Recherche pour le Développement (IRD), UNEP-WCMC, The World Fish Center, and WRI, 2011.

158. Briggs, J. C. 1974. *Marine Zoogeography.* New York: McGraw-Hill.

159. Allen, G. R., R. Steene, and M. Allen. 1998. *A Guide to Angelfishes and Butterflyfishes.* Perth, Australia: Odyssey Publishing/Tropical Reef Research.

160. Calculated at WRI based on data from LandScan High Resolution Global Population Data Set, Oak Ridge National Laboratory, 2007 and coral reef data from the Institute for Marine Remote Sensing, University of South Florida (IMaRS/USF), Institut de Recherche pour le Développement (IRD), UNEP-WCMC, The World Fish Center, and WRI, 2011.

161. Ateweberhan, M., and T. R. McClanahan. 2010. "Relationship between Historical Sea-Surface Temperature Variability and Climate Change-Induced Coral Mortality in the Western Indian Ocean." *Marine Pollution Bulletin* 60: 964–970.

162. Edwards, A. J., S. Clark, H. Zahir, A. Rajasuriya, A. Naseer, and J. Rubens. 2001. "Coral Bleaching and Mortality on Artificial and Natural Reefs in Maldives in 1998, Sea Surface Temperature Anomalies, and Initial Recovery." *Marine Pollution Bulletin* 42: 7–15.

163. Spencer, T., K. A. Teleki, C. Bradshaw, and M. D. Spalding. 2000. "Coral Bleaching in the Southern Seychelles During the 1997-1998 Indian Ocean Warm Event." *Marine Pollution Bulletin* 40: 569–586.

164. McClanahan, T. R., M. Ateweberhan, N. A. J. Graham, S. K. Wilson, C. R. Sebastián, M. M. M. Guillaume, and J. H. Bruggemann. 2007. "Western Indian Ocean Coral Communities: Bleaching Responses and Susceptibility to Extinction." *Marine Ecology Progress Series* 337: 1–13.

165. Wildlife Conservation Society. 2010. *Troubled Waters: Massive Coral Bleaching in Indonesia*. Accessible at: www.wcs.org/new-and-noteworthy/aceh-coral-bleaching.aspx. Accessed: September 2010.

166. Wilkinson, C., D. Souter, and J. Goldberg. 2005. *Status of Coral Reefs in Tsunami Affected Countries: 2005.* Townsville, Australia: Australian Institute of Marine Science.

167. Ministry of Planning and National Development, Republic of Maldives. 2008. *Analytical Report 2006: Population and Housing Census 2006.* Male, Republic of Maldives: Ministry of Planning and National Development.

168. Wilkinson, C. *Status of Coral Reefs of the World: 2008.* (Global Coral Reef Monitoring Network and Reef and Rainforest Research Centre, 2008).

169. Spalding, M. D. 2006. "Illegal Sea Cucumber Fisheries in the Chagos Archipelago." *SPC Beche-de-mer Information Bulletin* 23: 32–34.

170. Sheppard, C. R. C., A. Harris and A. L. S. Sheppard. Archipelago-Wide Coral Recovery Patterns since 1998 in the Chagos Archipelago, Central Indian Ocean. *Marine Ecology Progress Series* 362, 109-117 (2008).

171. Graham, N. A. J., S. K. Wilson, S. Jennings, N. V. C. Polunin, J. Robinson, J. P. Bijoux, and T. M. Daw. 2007. "Lag Effects in the Impacts of Mass Coral Bleaching on Coral Reef Fish, Fisheries, and Ecosystems." *Conservation Biology* 21: 1291–1300.

172. Graham, N. A. J. et al. 2008. "Climate Warming, Marine Protected Areas and the Ocean-Scale Integrity of Coral Reef Ecosystems." *PLoS ONE* 3: e3039.

173. McClanahan, T. R., N. A. J. Graham, J. M. Calnan, and M. A. MacNeil. 2007. "Toward Pristine Biomass: Reef Fish Recovery in Coral Reef Marine Protected Areas in Kenya." *Ecological Applications* 17: 1055–1067.

174. McClanahan, T. R. 2010. "Effects of Fisheries Closures and Gear Restrictions on Fishing Income in a Kenyan Coral Reef." *Conservation Biology* 24: 1519–1528.

175. De Santo, E. M., P. J. S. Jones, and A. M. M. Miller.2010. " Fortress Conservation at Sea: A Commentary on the Chagos Marine Protected Area." *Marine Policy* 35: 258–260; Mangi, S., T. Hooper, L. Rodwell, D. Simon, D. Snoxell, M. Spalding, and P. Williamson. 2010. "Establishing a Marine Protected Area in the Chagos Archipelago: Socio-Economic Considerations." Report of Workshop held January 7, 2010, Royal Holloway, University of London; Sand, P. H. 2010. "The Chagos Archipelago – Footprint of Empire, or World Heritage?" *Environmental Policy and Law* 40: 232–242.

176. Marine Research Center. 2009. *Maldives, the First Country in the Region to Ban Shark Fishing*. Accessible at: www.mrc.gov.mv/index.php/news_events/maldives_bans_shark_fishing. Accessed: September 13, 2010.

177. Briggs, J. C. 2005. "Coral Reefs: Conserving the Evolutionary Sources." *Biological Conservation* 126: 297–305; Carpenter, K. E., and V. G. Springer. 2005. "The Center of the Center of Marine Shore fish Biodiversity: The Philippine Islands." *Environmental Biology of Fishes* 72: 467–480.

178. Veron, J. E. N., L. M. Devantier, E. Turak, A. L. Green, S. Kininmonth, M. Stafford-Smith, and N. Peterson. 2009. "Delineating the Coral Triangle." *Galaxea, Journal of Coral Reef Studies* 11: 91-100.

179. Omori, M., K. Takahashi, N. Moriwake, K. Osada, T. Kimura, F. Kinoshita, S. Maso, K. Shimoike, and K. Hibino.2004. *Coral Reefs of Japan*. Tokyo, Japan: Ministry of Environment; Yamano, H., K. Hori, M. Yamauchi, O. Yamagawa, and A. Ohmura. 2001. "Highest-Latitude Coral Reef at Iki Island, Japan." *Coral Reefs* 20: 9–12.

180. Spalding, M. D., M. Kainuma and L. Collins. *World Atlas of Mangroves*. (Earthscan, with International Society for Mangrove Ecosystems, Food and Agriculture Organization of the United Nations, UNEP World Conservation Monitoring Centre, United Nations Scientific and Cultural Organisation, United Nations University, 2010).

181. Spalding, M. D., M. L. Taylor, C. Ravilious, F. T. Short, and E. P. Green. 2003. "Global Overview: The Distribution and Status of Seagrasses." In E. P. Green and F. T. Short, eds. *World Atlas of Seagrasses*. Berkeley, CA: University of California Press.

182. Nagelkerken, I. 2009. "Evaluation of Nursery Function of Mangroves and Seagrass Beds for Tropical Decapods and Reef Fishes: Patterns and Underlying Mechanisms." In Ivan Nagelkerken, ed. *Ecological Connectivity among Tropical Coastal Ecosystems*. New York: Springer.

183. Calculated at WRI based on data from LandScan High Resolution Global Population Data Set, Oak Ridge National Laboratory, 2007 and coral reef data from the Institute for Marine Remote Sensing, University of South Florida (IMaRS/USF), Institut de Recherche pour le Développement (IRD), UNEP-WCMC, The World Fish Center, and WRI, 2011.

184. FAO. 2009. "Food Balance Sheets." *FAOSTAT*. Accessible at: http://faostat.fao.org/.

185. FAO. 2007. *The World's Mangroves 1980-2005*. A Thematic Study Prepared in the Framework of the Global Forest Resources Assessment 2005. Rome: Forestry Department, Food and Agriculture Organization of the United Nations.

186. Reef Check Indonesia, The Nature Conservancy, and Wildlife Conservation Society. 2010. Press Release: "Global Mass Bleaching of Coral Reefs in 2010. Urgent Call to Action. " August 19, 2010.

187. Spencer, T., and M. D. Spalding.2005. "Coral Reefs of Southeast Asia: Controls, Patterns and Human Impacts." In A.Gupta, ed. *Physical Geography of Southeast Asia*. Oxford: Oxford University Press; Shear McCann, K. 2000. "The Diversity-Stability Debate." *Nature* 405: 228–233; Worm, B. et al.2006. "Impacts of Biodiversity Loss on Ocean Ecosystem Services." *Science* 314: 787–790.

188. Russ, G. R., and A. C. Alcala. 1996. "Do Marine Reserves Export Adult Fish Biomass? Evidence from Apo Island, Central Philippines." *Marine Ecology Progress Series* 132: 1–9.

189. Only some of the locally managed marine areas for the Philippines were included in our analysis, due to a lack of comprehensive spatial data. Since the protected areas are mostly small, they would not greatly affect our regional findings.

190. Calculated at WRI based on data from LandScan High Resolution Global Population Data Set, Oak Ridge National Laboratory, 2007 and coral reef data from the Institute for Marine Remote Sensing, University of South Florida (IMaRS/USF), Institut de Recherche pour le Développement (IRD), UNEP-WCMC, The World Fish Center, and WRI, 2011.

191. Access Economics. 2007. "Measuring the Economic and Financial Value of the Great Barrier Reef Marine Park, 2005-2006." Canberra, Australia: Access Economics.

192. Russ, G., A. Cheal, A. Dolman, M. Emslie, R. Evans, I. Miller, H. Sweatman, and D. Williamson. 2008. "Rapid Increase in Fish Numbers Follows Creation of World's Largest Marine Reserve Network." *Current Biology* 18: R514–R515; Raymundo, L. J., A. R. Halford, A. P. Maypa, and A. M. Kerr. 2009. "Functionally Diverse Reef-Fish Communities Ameliorate Coral Disease." *Proceedings of the National Academy of Sciences* 106: 17067–17070.

193. Great Barrier Reef Marine Park Authority. 2010. "Marine Shipping Incident: Great Barrier Reef Marine Park - Douglas Shoal." Information Sheet 3. Accessible at: www.gbrmpa.gov.au/corp_site/oil_spill_and_shipping_incidents/shen_neng_1_grounding.

194. Bainbridge, Z. T., J. E. Brodie, J. W. Faithful, D. A. Sydes, and S. E. Lewis. 2009. "Identifying the Land-Based Sources of Suspended Sediments, Nutrients and Pesticides Discharged to the Great Barrier Reef from the Tully–Murray Basin, Queensland, Australia." *Marine and Freshwater Research* 60: 1081–1090; Hutchings, P., D. Haynes, K. Goudkamp, and L. McCook. 2005. "Catchment to Reef: Water Quality Issues in the Great Barrier Reef Region--an Overview of Papers." *Marine Pollution Bulletin* 51: 3; De'ath, G., and K. E. Fabricius. 2008. *Water Quality of the Great Barrier Reef: Distributions, Effects on Reef Biota and Trigger Values for the Protection of Ecosystem Health.* Final Report to the Great Barrier Reef Marine Park Authority. Townsville, Australia: Australian Institute of Marine Science, Townsville; Devlin, M., P. Harkness, L. McKinna, and J. Waterhouse. 2010. *Mapping of Risk and Exposure of Great Barrier Reef Ecosystems to Anthropogenic Water Quality: A Review and Synthesis of Current Status.* Report to the Great Barrier Reef Marine Park Authority. Townsville, Australia: Australian Centre for Tropical Freshwater Research.

195. Jackson, J. B. C. et al. 2001. "Historical Overfishing and the Recent Collapse of Coastal Ecosystems." *Science* 293: 629–637.

196. Briggs, J. C. 2005. "The Marine East Indies: Diversity and Speciation." *Journal of Biogeography* 32: 1517–1522.

197. Allen, G. R., and D. R. Robertson. 1997."An Annotated Checklist of the Fishes of Clipperton Atoll, Tropical Eastern Pacific." *Revistas de Biologia Tropical* 45: 813–844; Glynn, P. W., and J. S. Ault. 2000. "A Biogeographic Analysis and Review of the Far Eastern Pacific Coral Reef Region." *Coral Reefs* 19: 1–23; León-Tejera, H., E. Serviere-Zaragoza, and J. González-González. 1996."Affinities of the Marine Flora of the Revillagigedo Islands, Mexico." *Hydrobiologia* 326–327: 159–168; Robertson, D. R., J. S. Grove, and J. E. McCosker. 2004. "Tropical Transpacific Shore Fishes." *Pacific Science* 58: 507–565; Spalding, M. D. et al. 2007. "Marine Ecoregions of the World: A Bioregionalization of Coast and Shelf Areas." *BioScience* 57: 573–583.

198. Calculated at WRI based on data from LandScan High Resolution Global Population Data Set, Oak Ridge National Laboratory, 2007 and coral reef data from the Institute for Marine Remote Sensing, University of South Florida (IMaRS/USF), Institut de Recherche pour le Développement (IRD), UNEP-WCMC, The World Fish Center, and WRI, 2011.

199. South, G. R., P. Skelton, J. Veitayaki, A. Resture, C. Carpenter, C. Pratt, and A. Lawedrau. 2004. *Pacific Islands, GIWA Regional Assessment 62.* Kalmar, Sweden: University of Kalmar on behalf of United Nations Environment Programme.

200. Church, J. A., N. J. White, and J. R. Hunter. 2006. "Sea-Level Rise at Tropical Pacific and Indian Ocean Islands." *Global and Planetary Change* 53: 155–168.

201. Scales, H., A. Balmford, M. Liu, Y. Sadovy, and A. Manica. 2006. "Keeping Bandits at Bay?" *Science* 313: 612–614; Sibert, J., J. Hampton, P. Kleiber, and M. Maunder. 2006. "Biomass, Size, and Trophic Status of Top Predators in the Pacific Ocean." *Science* 314: 1773–1776; Zeller, D., S. Booth, G. Davis, and D. Pauly.2007. "Re-Estimation of Small-Scale Fishery Catches for U.S. Flag-Associated Island Areas in the Western Pacific: The Last 50 Years." *Fishery Bulletin* 105: 266–277.

202. Goldberg, J. et al. 2008. "Status of Coral Reef Resources in Micronesia and American Samoa." In C. Wilkinson, ed. *Status of Coral Reefs of the World: 2008.* Townsville, Australia: Global Coral Reef Monitoring Network and Reef and Rainforest Research Centre; Vieux, C., B. Salvat, Y. Chancerelle, T. Kirata, T. Rongo, and E. Cameron. 2008. "Status of Coral Reefs in Polynesia Mana Node Countries: Cook Islands, French Polynesia, Niue, Kiribati, Tonga, Tokelau and Wallis and Futuna." In C. Wilkinson, ed. *Status of Coral Reefs of the World: 2008.* Townsville, Australia: Global Coral Reef Monitoring Network and Reef and Rainforest Research Centre.

203. Veron, J. E. N. 2000. *Corals of the World.* Townsville, Australia: Australian Institute of Marine Science.

204. Calculated at WRI based on data from LandScan High Resolution Global Population Data Set, Oak Ridge National Laboratory, 2007 and coral reef data from the Institute for Marine Remote Sensing, University of South Florida (IMaRS/USF), Institut de Recherche pour le Développement (IRD), UNEP-WCMC, The World Fish Center, and WRI, 2011.

205. Precht, W. F. 2002. "Endangered Acroporid Corals of the Caribbean." *Coral Reefs* 21: 41–42.

206. Miller, J., R. Waara, E. Muller, and C. Rogers. 2006. "Coral Bleaching and Disease Combine to Cause Extensive Mortality on Reefs in U.S. Virgin Islands." *Coral Reefs* 25: 418; Muller, E., C. Rogers, A. Spitzack, and R. van Woesik. 2008. "Bleaching Increases Likelihood of Disease on *Acropora Palmata* (Lamarck) in Hawksnest Bay, St John, U.S. Virgin Islands." *Coral Reefs* 27: 191–195.

207. Alvarez-Filip, L., N. K. Dulvy, J. A. Gill, I. M. Côté, and A. R. Watkinson. 2009. "Flattening of Caribbean Coral Reefs: Region-Wide Declines in Architectural Complexity." *Proceedings of the Royal Society B: Biological Sciences* 276: 3019–3025.

208. Pandolfi, J. M., J. B. C. Jackson, N. Baron, R. H. Bradbury, H. M. Guzman, T. P. Hughes, C. V. Kappel, F. Micheli, J. C. Ogden, H. P. Possingham, and E. Sala. 2005. "Are U.S. Coral Reefs on the Slippery Slope to Slime?" *Science* 307: 1725–1726.

209. Kikuchi, R. K. P., Z. M. A. N. Leao, V. Testa, L. X. C. Dutra, and S. Spano. 2003. "Rapid Assessment of the Abrolhos Reefs, Eastern Brazil (Part 1: Stony Corals and Algae)." *Atoll Research Bulletin* 496: 172–187.

210. Mumby, P. J. and A. R. Harborne. 2010. "Marine Reserves Enhance the Recovery of Corals on Caribbean Reefs." *PLoS ONE* 5: e8657.

211. Appeldoorn, R. S., and K. C. Lineman. 2003. "A Caribbean-Wide Survey of Marine Reserves:–Spatial Coverage and Attributes of Effectiveness." *Gulf and Caribbean Research* 14: 139-154.

212. Moberg, F., and C. Folke.1999. "Ecological Goods and Services of Coral Reef Ecosystems." *Ecological Economics* 29: 215–233.

213. Whittingham, E., J. Campbell, and P. Townsley. 2003. *Poverty and Reefs. Volume 1: A Global Overview.* Paris, France: DFID–IMM–IOC/UNESCO.

214. Reef territories that are only inhabited by military or scientific personnel are not included.

215. UN-OHRLLS. 2006. *List of Least Developed Countries.* Accessible at: www.un.org/special-rep/ohrlls/ldc/list.htm. Accessed: July 20, 2009.

216. Bettencourt, S. et al. 2006. *Not If but When: Adapting to Natural Hazards in the Pacific Islands Region.* A Policy Note. Washington, DC: The World Bank; Briguglio, L. 1995. "Small Island Developing States and Their Economic Vulnerabilities." *World Development* 23: 1615–1632; Pelling, M., and J. I. Uitto. 2001. "Small Island Developing States: Natural Disaster Vulnerability and Global Change." *Environmental Hazards* 3: 49–62.

217. Allison, E. H., A. L. Perry, M.-C. Badjeck, W. N. Adger, K. Brown, D. Conway, A. S. Halls, G. M. Pilling, J. D. Reynolds, N. L. Andrew, and N. K. Dulvy. 2008. "Vulnerability of National Economy to the Impacts of Climate Change on Fisheries." *Fish and Fisheries* 10: 173–196; IPCC. 2001. *Climate Change 2001: Impacts, Adaptation, and Vulnerability. Contribution of Working Group II to the Third Assessment Report of the Intergovernmental Panel on Climate Change.* Cambridge: Cambridge University Press; Turner, B. L., II et al. 2003. "A Framework for Vulnerability Analysis in Sustainability Science." *Proceedings of the National Academy of Sciences* 100: 8074–8079.

218. The exposure component combines the Reefs at Risk integrated threat index with the ratio of reef to land area for each country and territory in the assessment. See online for further details.

219. Salvat, B. 1992. "Coral Reefs - a Challenging Ecosystem for Human Societies." *Global Environmental Change* 2: 12–18; Wilkinson, C. R. 1996. "Global Change and Coral Reefs: Impacts on Reefs, Economies and Human Cultures." *Global Change Biology* 2: 547–558.

220. Loper, C. et al. 2008. *Socioeconomic Conditions Along the World's Tropical Coasts: 2008.* Silver Spring, MD: National Oceanic and Atmospheric Administration, Global Coral Reef Monitoring Network, and Conservation International.

221. Turner, R. A., A. Cakacaka, N. A. J. Graham, N. V. C. Polunin, M. S. Pratchett, S. M. Stead, and S. K. Wilson. 2007. "Declining Reliance on Marine Resources in Remote South Pacific Societies: Ecological Versus Socio-Economic Drivers." *Coral Reefs* 27: 997–1008.

222. A multiplicative index of vulnerability was selected over an additive (averaging) model. Results from the two models were highly correlated (r=0.84). However, the multiplicative model was selected because it tended to award high vulnerability ratings on the basis of high scores for all three components (i.e., exposure, reef dependence, adaptive capacity). In contrast, the additive model produced some very high vulnerability scores that were driven by a very high score on a single component.

223. Calculated at WRI based on population data from LandScan High Resolution Global Population Data Set, Oak Ridge National Laboratory, 2007; and coral reef data from the Institute for Marine Remote Sensing, University of South Florida (IMaRS/USF), Institut de Recherche pour le Développement (IRD), UNEP-WCMC, The World Fish Center, and WRI, 2011.

224. Aswani, S., and I. Vaccaro.2008. "Lagoon Ecology and Social Strategies: Habitat Diversity and Ethnobiology." *Human Ecology* 36: 325–341; Chapman, M. D.1987. "Women's Fishing in Oceania." *Human Ecology* 15: 267–288.

225. Hoon, V. 2003. "A Case Study from Lakshadweep." In Emma Whittingham, Jock Campbell, and Philip Townsley, eds. *Poverty and Reefs.* Vol. II: Case studies. Paris, France: DFID-IMM-IOC/UNESCO.

226. Gillett, R. 2009. *The Contribution of Fisheries to the Economies of Pacific Island Countries and Territories.* Manila: Asian Development Bank.

227. Estimates included full-time, part-time, commercial, and subsistence fishers. Where reef-specific data were not available, estimates were derived by combining the number of small-scale coastal fishers in reef regions with the proportion of coastal population within 30 km of reefs. See online for further details.

228. The top quartile = 27 countries and territories.

229. Data are from FAO food balance sheets, national Household Income and Expenditure Surveys, and other studies. Consumption includes marine and freshwater fish and invertebrates. Further details are available online.

230. Note that these estimates describe fish and seafood consumption at the national scale. Local-scale consumption is likely to be higher than national estimates in many reef regions, particularly where countries have significant inland populations. Estimates include all sources of fish and seafood (including inland and canned supplies), and are therefore less likely to be representative of reef fishery consumption nations and territories with significant industrial and/or inland fisheries.

231. Bell, J. D., M. Kronen, A. Vunisea, W. J. Nash, G. Keeble, A. Demmke, S. Pontifex, and S. Andréfouët. 2008. "Planning the Use of Fish for Food Security in the Pacific." *Marine Policy* 33: 64–76.

232. Census and Statistics Department, Government of Hong Kong Special Administrative Region.

233. Chávez, E. A. 2009. "Potential Production of the Caribbean Spiny Lobster (Decapoda, Palinura) Fisheries." *Crustaceana* 82: 1393–1412.

234. The values of most exports of dead reef fish and invertebrates for food are particularly difficult to estimate, because export statistics typically distinguish these items by product (e.g. "fish fillets, frozen"), rather than by species.

235. It is assumed that where these reef products are exported, they are indicative of trade routes that other reef commodities are likely to follow.

236. Based on countries with registered dive centers.

237. Based on tourism receipts and current GDP.

238. The number of registered dive centers were divided by annual tourist arrivals, and then multiplied by the value of annual tourist receipts as a proportion of GDP.

239. Tourism Corporation of Bonaire (TCB). 2009. *Bonaire Tourism: Annual Statistics Report 2008*. Kralendijk, Bonaire: Tourism Corporation of Bonaire.

240. Brander, R. W., P. Kench, and D. Hart. 2004. "Spatial and Temporal Variations in Wave Characteristics across a Reef Platform, Warraber Island, Torres Strait, Australia." *Marine Geology* 207: 169–184.

241. See technical notes at www.wri.org/reefs.

242. Calculated at WRI based on population data from LandScan High Resolution Global Population Data Set, Oak Ridge National Laboratory, 2007; coastline data from the National Geospatial Intelligence Agency, World Vector Shoreline, 2004; and coral reef data from the Institute for Marine Remote Sensing, University of South Florida (IMaRS/USF), Institut de Recherche pour le Développement (IRD), UNEP-WCMC, The World Fish Center, and WRI, 2011.

243. Australian Institute of Marine Science and Maldives Marine Research Centre. 2005. *An Assessment of Damage to Maldivian Coral Reefs and Baitfish Populations from the Indian Ocean Tsunami*. Canberra, Australia: Commonwealth of Australia; Ministry of Tourism. 2009. *Tourism Yearbook 2009*. Male: Republic of Maldives.

244. Smit, B., and J. Wandel. 2006. "Adaptation, Adaptive Capacity, and Vulnerability." *Global Environmental Change* 16: 282–292.

245. See technical notes at www.wri.org/reefs.

246. Per capita GDP, derived from purchasing power parity (PPP) methods, which account for differences in the relative prices of goods and services among nations.

247. Connell, J., and R. P. C. Brown. 2005. *Remittances in the Pacific: An Overview*. Manila: Asian Development Bank; The World Bank. 2008. *Migration and Remittances Factbook 2008*. Washington, DC: World Bank.

248. The Worldwide Governance Indicators project of the World Bank (http://info.worldbank.org/governance/wgi/index.asp) reports national-scale indicators of six dimensions of governance: Voice and Accountability, Political Stability and Absence of Violence, Government Effectiveness, Regulatory Quality, Rule of Law, and Control of Corruption. For this analysis, the six components were averaged to yield an integrated governance index.

249. Sea Around Us Project. 2010. *Fisheries, Ecosystems and Biodiversity*. Accessible at: www.seaaroundus.org/data/. Accessed: July 2009; Sumaila, U. R., and D. Pauly. 2006. *Catching More Bait: A Bottom-up Re-Estimation of Global Fisheries Subsidies (2nd Version). Fisheries Centre Research Reports*, Vol. 14. Vancouver: Fisheries Centre, University of British Columbia.

250. The value of subsidies are assessed relative to the value of fisheries landings.

251. Agricultural land availability is assessed as agricultural land per agricultural worker.

252. Allison, E. H., and F. Ellis. 2001. "The Livelihoods Approach and Management of Small-Scale Fisheries." *Marine Policy* 25: 377–388.

253. Schwarz, A., C. Ramofafia, G. Bennett, D. Notere, A. Tewfik, C. Oengpepa, B. Manele, and N. Kere. 2007. *After the Earthquake: An Assessment of the Impact of the Earthquake and Tsunami on Fisheries-Related Livelihoods in Coastal Communities of Western Province, Solomon Islands*. Honiara, Solomon Islands: WorldFish Center and WWF-Solomon Islands Programme.

254. Threat levels for reefs in Bermuda range from medium to very high. However, the value for the exposure index is very high because this component combines threat levels and the ratio of reef to land area (which is very high for Bermuda).

255. The World Bank. 2010. *The World Bank: Country Data for Jamaica*. Accessible at: http://data.worldbank.org/country/jamaica . Accessed: July 2010.

256. Hardt, M. J. 2009. "Lessons from the Past: The Collapse of Jamaican Coral Reefs." *Fish and Fisheries* 10: 143–158; Hughes, T. P. 1994. "Catastrophes, Phase Shifts, and Large-Scale Degradation of a Caribbean Coral Reef." *Science* 265: 1547–1551.

257. Sauni, S. et al. 2007. *Nauru Country Report: Profile and Results from in-Country Survey Work*. Noumea, New Caledonia: Secretariat of the Pacific Community.

258. Ireland, C., D. Malleret, and L. Baker. 2004. *Alternative Sustainable Livelihoods for Coastal Communities: A Review of Experience and Guide to Best Practice*. Nairobi: IUCN.

259. Cattermoul, B., P. Townsley, and J. Campbell. 2008. *Sustainable Livelihoods Enhancement and Diversification (SLED): A Manual for Practitioners*. Gland, Switzerland: IUCN, International Union for Conservation of nature.

260. Non-use values, such as "existence value" and "bequest value" are less frequently estimated than the "use values" described above. Non-use values are frequently large values with wide error bounds, but can also help to inform policy.

261. European Communities. 2008. *The Economics of Ecosystems and Biodiversity (*TEEB Report). Wesseling, Germany: European Communities.

262. Munro, J. L. 1974. "The Biology, Ecology, Exploitation and Management of Caribbean Reef Fishes." Kingston, Jamaica: Zoology Department of the University of the West Indies; McAllister, D. E. 1988. "Environmental, Economic and Social Costs of Coral Reef Destruction in the Phillipines." *Galaxea* 7: 161–178.

263. Calculated at WRI based on population data from LandScan High Resolution Global Population Data Set, Oak Ridge National Laboratory, 2007; coastline data from the National Geospatial Intelligence Agency, World Vector Shoreline, 2004; and coral reef data from the Institute for Marine Remote Sensing, University of South Florida (IMaRS/USF), Institut de Recherche pour le Développement (IRD), UNEP-WCMC, The World Fish Center and WRI, 2011.

264. Burke, L. and J. Maidens. 2004. *Reefs at Risk in the Caribbean.* Washington, DC: World Resources Institute.

265. Hoegh-Guldberg, O., and H. Hoegh-Guldberg. 2004. *Implications of Climate Change for Australia's Great Barrier Reef.* Sydney, Australia: World Wildlife Fund.

266. Burke, L., E. Selig, and M. Spalding. 2002. *Reefs at Risk in Southeast Asia.* Washington, DC: World Resources Institute.

267. van Beukering, P., L. Brander, E. Tompkins, and E. McKenzie. 2007. *Valuing the Environment in Small Islands: An Environmental Economics Toolkit.* Peterborough, UK : Joint Nature Conservation Commitee.

268. Cooper, E., L. Burke, and N. Bood. 2008. *Coastal Capital: Belize the Economic Contribution of Belize's Coral Reefs and Mangroves.* Washington, DC: World Resource Institute; Attorney General of Belize vs. MS Westerhaven Schiffahrts GMBH and Co KG and Reider Shipping BV. Supreme Court of Belize, April 26, 2010.

269. Dixon, A., L. Fallon Scura, and T. Van't Hof. 1993. "Meeting Ecological and Economic Goals: Marine Parks in the Caribbean." *Ambio* 22: 117-125.

270. IUCN defines a protected area as "a clearly defined geographical space, recognized, dedicated and managed through legal or effective means, to achieve the long-term conservation of nature with associated ecosystem services and cultural values." For "marine" it includes any site with subtidal or intertidal waters.

271. Selig, E. R., and J. F. Bruno. 2010. "A Global Analysis of the Effectiveness of Marine Protected Areas in Preventing Coral Loss." *PLoS ONE* 5: 7.

272. Jones, P. 2007. "Point-of-View: Arguments for Conventional Fisheries Management and against No-Take Marine Protected Areas: Only Half of the Story?" *Reviews in Fish Biology and Fisheries* 17: 31–43.

273. Bartlett, C. Y., K. Pakoa, and C. Manua. 2009."Marine Reserve Phenomenon in the Pacific Islands." *Marine Policy* 33: 673-678.

274. McClanahan, T. R., M. J. Marnane, J. E. Cinner, and W. E. Kiene. 2006. "A Comparison of Marine Protected Areas and Alternative Approaches to Coral-Reef Management." *Current Biology* 16: 1408–1413.

275. Fernandes, L. et al. 2005. "Establishing Representative No-Take Areas in the Great Barrier Reef: Large-Scale Implementation of Theory on Marine Protected Areas." *Conservation Biology* 19: 1733–1744.

276. Govan, H. 2009. *Status and Potential of Locally-Managed Marine Areas in the South Pacific: Meeting Nature Conservation and Sustainable Livelihood Targets through Wide-Spread Implementation of LMMAs.* Suva, Fiji: Coral Reef Initiatives for the Pacific, with SPREP/WWF/WorldFish-Reefbase.

277. Govan, H. 2009. "Achieving the Potential of Locally Managed Marine Areas in the South Pacific." *SPC Traditional Marine Resource Management and Knowledge Information Bulletin* 25: 16–25.

278. Alcala, A. C., and G. R. Russ. 2006. "No-Take Marine Reserves and Reef Fisheries Management in the Philippines: A New People Power Revolution." *Ambio* 35: 245–254.

279. Bartlett, C. Y., T. Maltali, G. Petro, and P. Valentine.2010. "Policy Implications of Protected Area Discourse in the Pacific Islands." *Marine Policy* 34: 99–104; Gell, F. 2002. "Success in Soufrière. The Soufrière Marine Mangement Area, St Lucia: A Community Initiative That Has Worked for Fishers." *Reef Encounter, Newsletter of the International Society for Reef Studies* 31: 30–32; Johannes, R. E. 2002. "The Renaissance of Community-Based Marine Resource Management in Oceania." *Annual Review of Ecology and Systematics* 33: 317–340.

280. Hughes, T. P., D. R. Bellwood, C. Folke, R. S. Steneck, and J. Wilson. 2005. "New Paradigms for Supporting the Resilience of Marine Ecosystems." *Trends in Ecology and Evolution* 20: 380–386; Klein, C. J., C. Steinback, M. Watts, A. J. Scholz, and H. P. Possingham. 2010. "Spatial Marine Zoning for Fisheries and Conservation." *Frontiers in Ecology and the Environment* 8: 349–353; Pauly, D., V. Christensen, S. Guenette, T. J. Pitcher, U. R. Sumaila, C. J. Walters, R. Watson, and D. Zeller. 2002. "Towards Sustainability in World Fisheries." *Nature* 418: 689–695; Steneck, R. S., C. B. Paris, S. N. Arnold, M. C. Ablan-Lagman, A. C. Alcala, M. J. Butler, L. J. McCook, G. R. Russ, and P. F. Sale. 2009. "Thinking and Managing Outside the Box: Coalescing Connectivity Networks to Build Region-Wide Resilience in Coral Reef Ecosystems." *Coral Reefs* 28: 367; The Nature Conservancy. 2010. *R2 Reef Resilience: Building Resilience into Coral Reef Conservation.* Accessible at: www.reefresilience.org. Accessed: May 28,2010; Toropova, C., I. Meliane, D. Laffoley, E. Matthews, and M. Spalding. 2010. *Global Ocean Protection: Present Status and Future Possibilities.* Cambridge, UK, Arlington, VA, Tokyo, New York, Gland, Switzerland, and Washington, DC: IUCN WCPA,UNEP-WCMC, TNC, UNU, WCS.

281. IUCN-WCPA. 2008. *Establishing Marine Protected Area Networks - Making It Happen.* Gland, Switzerland, Washington, DC, and Arlington, VA: IUCN-WCPA, National Oceanic and Atomospheric Association, The Nature Conservancy; Lowry, G. K., A. T. White, and P. Christie. 2009. "Scaling up to Networks of Marine Protected Areas in the Philippines: Biophysical, Legal, Institutional, and Social Considerations." *Coastal Management* 37: 274–290; UNEP-WCMC. 2008. *National and Regional Networks of Marine Protected Areas: A Review of Progress.* Cambridge, UK: UNEP-WCMC.

282. For *Reefs at Risk Revisited*, we compiled a new global dataset of MPAs near coral reefs. Our definition of a coral reef MPA includes all sites that overlap with coral reefs on the map (1,712 sites), but also those that are known (from a variety of sources) to contain reefs. To these we added a third category—sites considered likely to contain reefs or reef species. These are the sites with offshore or subtidal areas that occur within 20 km of a coral reef. We included these sites to avoid missing key MPAs due to mapping errors or inaccuracies. For example, we lack accurate boundary information for some MPAs, while reef maps themselves are also missing some areas of reef (notably small isolated patches or coral communities that are too small or deep to be properly mapped). The primary source for this information is the World Database of Protected Areas (WDPA), which provided the majority of sites. In addition, Reef Base provided information on over 600 LMMAs for Pacific Islands and in the Phillipines. The Nature Conservancy provided data on over 100 additional sites in Indonesia, while reviewers provided about 50 additional sites. For the analysis, we differentiated the nine different management zones within the Great Barrier Reef Marine Park. The combined areas in each zone are substantial, and each zone offers strikingly different levels of protection. The final total of 2,679 sites is undoubtedly the most comprehensive listing ever produced. While our estimates of total reef area protected are derived from those sites which directly overlap our reef map, it is likely that we have an accurate picture of overall protection as these include all of the larger coral reef MPAs.

283. A number of studies have attempted to develop tools for assessing "management effectiveness," although to date such measures have only been applied to a small proportion of sites. These include: Hocking, M., D. Stolton, and N. Dudley. 2000. *Evaluating Effectiveness: a Framework for Assessing the Management of Protected Areas*. Gland, Switzerland: IUCN-World Conservation Union; Pomeroy, R.S., J..E. Parks, and L.M. Watson. 2004. *How is you MPA doing? A guidebook of natural and social indicators for evaluating marine protected areas management effectiveness*. Gland, Switzerland and Camridge, UK: IUCN, WWF and NOAA.

284. Unlike some broader measures of management effectiveness, our primary interest was in ecological effectiveness, and given the challenges in any such survey, we reduced our focus simply to the consideration of the impact of an MPA on the threat of overfishing. Building on earlier work undertaken in the *Reefs at Risk in the Caribbean* and *Reefs at Risk in Southeast Asia* analyses, and using input from a number of other experts and literature review, sites were scored using a 3-point scale: 1) Effective, where the site is managed sufficiently well that *in situ* threats are not undermining natural ecosystem function; 2) Partially effective, where the site is managed such that *in situ* threats are significantly lower than adjacent non-managed sites, but there may still be some detrimental effects on ecosystem function; and 3) Ineffective, where the site is unmanaged, or management is unsufficient to reduce *in situ* threats in any meaningful way. Given that the sampling drew on field knowledge by regional experts rather than field practitioners, there is likely to be a sampling bias toward better-known sites, with perhaps a higher proportion of effective sites than would be found overall.

285. PISCO. 2008. *The Science of Marine Reserves. Second Edition: Latin America and the Caribbean*. Santa Barbara, CA: The Partnership for Interdisciplinary Studies of Coastal Oceans; Russ, G., A. Cheal, A. Dolman, M. Emslie, R. Evans, I. Miller, H. Sweatman, and D. Williamson. 2008. "Rapid Increase in Fish Numbers Follows Creation of World's Largest Marine Reserve Network." *Current Biology* 18: R514–R515.

286. Of the 1,536 sites that the 2008 studies reviewed (see Spalding, et al. and Wood, et al.), only 30 percent were entirely or partially no-take, and this included many sites outside of coral reef areas.

287. Spalding, M. D., L. Fish and L. J. Wood. 2008. "Toward Representative Protection of the World's Coasts and Oceans: Progress, Gaps, and Opportunities." *Conservation Letters* 1: 217–226; Wood, L. J., L. Fish, J. Laughren, and D. Pauly. 2008. "Assessing Progress Towards Global Marine Protection Targets: Shortfalls in Information and Action." *Oryx* 42: 340–351.

288. Two very large MPAs, the Papahānaumokuākea Marine National Monument in Hawaii and the Chagos Marine Protected Area, are not included in the assessment of no-take area. Both have some allowance for low levels of fishing, although the actual fishing pressure remains very low, regardless. Papahānaumokuākea is expected to become fully no-take in mid-2011. If these sites were included, they would add over 4,000 sq km to these no-take statistics.

289. Obura, D. 2005. "Resilience and Climate Change: Lessons from Coral Reefs and Bleaching in the Western Indian Ocean." *Estuarine, Coastal and Shelf Science* 63: 353–372.

290. PISCO. 2008. *The Science of Marine Reserves. Second Edition: Latin America and the Caribbean*. Santa Barbara, CA: The Partnership for Interdisciplinary Studies of Coastal Oceans.

About the Authors

Lauretta Burke is a Senior Associate in WRI's People and Ecosystems Program. She has an M.A. in Environment and Resource Policy from the George Washington University and an M.A. in Geography from the University of California, Santa Barbara. Lauretta leads WRI's work on coastal ecosystems, including the *Reefs at Risk* project and *Coastal Capital* series on valuation of coral reefs.

Kathleen Reytar is a Research Associate in WRI's People and Ecosystems Program. She has a Master of Environmental Science & Management from the Bren School at the University of California, Santa Barbara and a B.S. in Civil and Environmental Engineering from the Johns Hopkins University. Katie specializes in geographic information systems and coastal marine resources management.

Mark Spalding is a Senior Marine Scientist with The Nature Conservancy's Global Marine Team and works out of the Department of Zoology at the University of Cambridge. He has authored many books and papers on the distribution, condition, and conservation of marine ecosystems, including the *World Atlas of Mangroves*, *The World's Protected Areas*, *World Atlas of Coral Reefs* and earlier *Reefs at Risk* studies.

Allison Perry was a Postdoctoral Fellow with The WorldFish Center at the time of writing. Her research has addressed a wide range of issues related to human pressure on marine ecosystems, including large-scale effects of climate change on coral reefs; vulnerability of fishing-dependent economies to climate change; trade and conservation of threatened marine fishes; and reef dependence. Allison has recently joined Oceana in Madrid.

Notes

PHOTO: DAVID BURDICK

Notes

PHOTO: MARY LOU FROST